大学计算机规划教材

Java
大学实用教程学习指导
（第3版）

◆ 张跃平　耿祥义　雷金娥　编著

电子工业出版社
Publishing House of Electronics Industry
北京·BEIJING

内 容 简 介

本书是《Java 大学实用教程（第 3 版）》（ISBN 978-7-121-14913-9）的配套学习指导书，除了按照主教材的章节配备实验指导外，还增加了一个综合实验——走迷宫游戏。

每章的实验指导由知识概括、实验内容和知识扩展三部分内容组成，学生可按照实验的要求上机编写程序。每个实验都提供了程序模板，学生完成实验后需填写实验报告。知识扩展是对实验内容的一个补充，结合实例讲解主教材未涉及的一些知识或已学知识的深入讨论。综合实验的目的是综合运用所学知识设计一个完整软件。

本书适合作为高等院校面向对象程序设计（Java）相关课程的学习参考书，也可供自学者参考。

未经许可，不得以任何方式复制或抄袭本书之部分或全部内容。
版权所有，侵权必究。

图书在版编目(CIP)数据

Java 大学实用教程学习指导 / 张跃平，耿祥义，雷金娥编著. —3 版. —北京：电子工业出版社，2012.8
大学计算机规划教材
ISBN 978-7-121-17314-1

Ⅰ. ①J… Ⅱ. ①张… ②耿… ③雷 Ⅲ. ①JAVA 语言－程序设计－高等学校－教材 Ⅳ. ①TP312
中国版本图书馆 CIP 数据核字（2012）第 121243 号

策划编辑：章海涛
责任编辑：章海涛　　　　　特约编辑：曹剑锋
印　　刷：北京虎彩文化传播有限公司
装　　订：北京虎彩文化传播有限公司
出版发行：电子工业出版社
　　　　　北京市海淀区万寿路 173 信箱　邮编　100036
开　　本：787×1092　1/16　印张：12.5　字数：340 千字
版　　次：2005 年 3 月第 1 版
　　　　　2012 年 8 月第 3 版
印　　次：2022 年 1 月第 9 次印刷
定　　价：28.00 元

凡所购买电子工业出版社图书有缺损问题，请向购买书店调换。若书店售缺，请与本社发行部联系，联系及邮购电话：(010) 88254888。
质量投诉请发邮件至 zlts@phei.com.cn，盗版侵权举报请发邮件至 dbqq@phei.com.cn。
服务热线：(010) 88258888。

第 3 版前言

本书是主教材《Java 大学实用教程（第 3 版）》（ISBN 978-7-121-14913-9）的配套学习指导书，目的是通过一系列实验练习使学生巩固所学的知识。

相对于第 2 版，本书修改了部分实验的内容，并增加了一些新的实验内容，特别是第 14 章的综合实验由原来的扫雷游戏更新为了走迷宫游戏。

每章由以下三部分组成。

1. 知识概括

这一部分总结了主教材相应章节的重点和难点知识。

2. 实验练习

这一部分由若干实验组成，每个实验主要包括五部分：

- 实验目的——让学生了解实验需要掌握哪些知识，实验将以这些知识为中心。
- 实验要求——该实验需要达到的基本标准。
- 程序模板——一个 Java 源程序，其中隐藏了需要学生重点掌握的代码，这部分代码要求学生来完成。模板起到引导作用，学生通过完成模板可以深入了解解决问题的方式。
- 实验指导与检查——针对实验的难点给出必要的提示，并要求学生向指导老师演示模板程序的运行效果。
- 实验报告——其中包括根据实验提出的一些问题或要求学生进一步编写的代码。对于实验报告中提出的问题，学生可能需要编写一些程序代码，才能给出一个正确的答案；对于要求学生编写的代码，学生必须按照要求编写。学生须完成该实验报告的填写，并由指导老师签字。

3. 知识扩展

这一部分是对主教材对应章节的知识的补充，结合实例讲解主教材未能涉及的一些知识或对已学知识的深入讨论。

读者可以登录到**华信教育资源网**（http://www.hxedu.com.cn）**下载实验用程序模板的完整源程序**，也可发邮件至 unicode@phei.com.cn 咨询。

作　者

作者简介

张跃平，现任大连交通大学副教授，具有多年从事 Java 语言教学经验。

耿祥义，1995 年中国科学技术大学博士毕业，获理学博士学位。1997 年从中山大学博士后流动站出站。现任大连交通大学教授，具有多年从事 Java 语言教学经验，已编写出版多部教材。

目 录

第 1 章 Java 语言概述 ··· 1
1.1 知识概述 ·· 2
1.2 实验练习 ·· 2
    1.2.1 一个简单的应用程序 ··· 2
    1.2.2 源文件的命名规则 ··· 4
1.3 知识扩展——联合编译 ·· 5

第 2 章 基本数据类型和数组 ··· 6
2.1 知识概述 ·· 7
2.2 实验练习 ·· 7
    2.2.1 输出俄文字母表 ·· 7
    2.2.2 从键盘输入数据 ·· 8
2.3 知识扩展——数组的快速复制 ··· 9

第 3 章 运算符、表达式和语句 ·· 12
3.1 知识概述 ··· 13
3.2 实验练习 ··· 13
    3.2.1 计算电费 ·· 13
    3.2.2 猜数字 ··· 14
3.3 知识扩展——使用 Arrays 类实现数组排序 ··· 15

第 4 章 类和对象 ··· 17
4.1 知识概述 ··· 18
4.2 实验练习 ··· 18
    4.2.1 机动车的类封装 ··· 18
    4.2.2 有理数的类封装 ··· 20
    4.2.3 公司和职员 ··· 23
    4.2.4 实例成员和类成员 ·· 25
    4.2.5 package 语句和 import 语句 ··· 27
4.3 知识扩展——Class 类的使用 ·· 29

第 5 章 继承和接口 ·· 32
5.1 知识概述 ··· 33
5.2 实验练习 ··· 33
    5.2.1 继承 ·· 33
    5.2.2 上转型对象 ··· 36
    5.2.3 接口回调 ·· 38
    5.2.4 异常处理 ·· 40
5.3 知识扩展——可变参数和断言语句 ··· 42

第 6 章 字符串和正则表达式 ·· 44
6.1 知识概述 ··· 45

6.2	实验练习	45
	6.2.1 String 类的常用方法	45
	6.2.2 StringBuffer 类的常用方法	47
	6.2.3 Scanner 类与字符串分解	49
	6.2.4 模式匹配	50
6.3	知识扩展——元词和定位元字符	51

第 7 章 常用实用类   54

7.1	知识概述	55
7.2	实验练习	55
	7.2.1 比较日期的大小	55
	7.2.2 随机布雷	56
	7.2.3 使用 TreeSet 排序	59
	7.2.4 使用 TreeMap 排序	60
7.3	知识扩展——排序和查找、自动装箱和自动拆箱	62

第 8 章 多线程   67

8.1	知识概述	68
8.2	实验练习	69
	8.2.1 使用 Thread 的子类创建线程	69
	8.2.2 使用 Thread 类创建线程	71
	8.2.3 吵醒休眠的线程	73
	8.2.4 排队买票	75
	8.2.5 线程联合	78
8.3	知识扩展——Timer 类和 TimerTask 类	80

第 9 章 输入流和输出流   83

9.1	知识概述	84
9.2	实验练习	85
	9.2.1 文件加密	85
	9.2.2 分析成绩单	86
	9.2.3 文件读取和模式匹配	88
	9.2.4 读/写基本类型数据	90
	9.2.5 对象的写入和读取	91
	9.2.6 使用 Scanner 解析文件	93
9.3	知识扩展——ZIP 文件的读取和制作	95

第 10 章 图形用户界面设计   98

10.1	知识概述	99
10.2	实验练习	99
	10.2.1 布局与日历	99
	10.2.2 猜数字游戏	103
	10.2.3 算术测试	105
	10.2.4 单词统计和排序	109
	10.2.5 华容道游戏	112
	10.2.6 字体对话框	118

	10.3	知识扩展——计时器	120
第 11 章		Java 中的网络编程	123
	11.1	知识概述	124
	11.2	实验练习	124
		11.2.1 读取服务器中的文件	124
		11.2.2 过滤网页中的内容	126
		11.2.3 使用套接字传输数据	130
		11.2.4 基于 UDP 的图像传输	134
	11.3	知识扩展——网络中的数据压缩和传输	138
第 12 章		Java 数据库操作	142
	12.1	知识概述	143
	12.2	实验练习	143
		12.2.1 JDBC-ODBC 桥接器	143
		12.2.2 查询、更新和插入操作	145
		12.2.3 预处理语句	147
		12.2.4 事务处理	149
	12.3	知识扩展——MySQL 简介	152
第 13 章		Java Applet	156
	13.1	知识概述	157
	13.2	实验练习	157
		13.2.1 播放音频	157
		13.2.2 绘制五角星	160
		13.2.3 左手画圆右手画方	161
		13.2.4 图像渐变	163
		13.2.5 读取服务器端文件	164
	13.3	知识扩展——Java 2D 简介	166
第 14 章		综合实验——走迷宫	170
	14.1	设计要求	171
	14.2	总体设计	171
	14.3	详细设计	172
		14.3.1 编写迷宫文件	172
		14.3.2 MazeWindow 类	173
		14.3.3 Maze 类	177
		14.3.4 WallOrRoad 类	181
		14.3.5 MazePoint 类	184
		14.3.6 PersonInMaze 类	186
		14.3.7 HandleMove 类	187
		14.3.8 所需图像	191
	14.4	代码调试	191
	14.5	软件发布	191
	14.6	实验后的练习	191

第 1 章　Java 语言概述

本章导读

- 知识概述
- 实验1　一个简单的应用程序
- 实验2　源文件的命名规则
- 知识扩展——联合编译

1.1 知识概述

Java 语言的出现源于对独立于平台语言的需要,即这种语言编写的程序不会因为芯片的变化而无法运行或运行错误。目前,随着网络的迅速发展,Java 语言的优势愈加明显,Java 已经成为网络时代最重要的语言之一。

本章要求读者初步了解 Java 的一些特点,如面向对象、多线程、动态、平台无关等,许多特点必须经过进一步的学习才能深入理解。Java 有三个重要平台:Java EE、Java SE 和 Java ME,分别针对大型服务器程序、一般应用程序和嵌入式程序的设计开发平台。Java SE 平台是学习掌握 Java 语言的最佳平台,而掌握 Java SE 又是进一步学习 Java EE 和 Java ME 所需的。本章要求读者重点掌握开发 Java 应用程序的基本步骤。

1.2 实验练习

1.2.1 一个简单的应用程序

1. 实验目的

本实验的目的是让学生掌握开发 Java 应用程序的三个步骤:编写源文件、编译源文件和运行应用程序。

图 1-1 简单的应用程序

2. 实验要求

编写一个简单的 Java 应用程序,该程序在命令行窗口中输出两行文字:"你好,很高兴学习 Java"和"We are students"。

3. 程序效果示例

程序效果如图 1-1 所示。

4. 程序模板

按模板要求,将【代码】[①]替换为程序代码。

Hello.java

```
public class Hello {
    public static void main (String args[]){
        【代码1】              //命令行窗口输出"你好,很高兴学习 Java"
        A a=new A();
        a.fA();
    }
}
class A {
    void fA() {
        【代码2】              //命令行窗口输出"We are students"
    }
}
```

[①] 【代码】泛指程序段中的【代码1】、【代码2】等,以下同。

5. 实验指导与检查

步骤 1：打开一个文本编辑器。如果是 Windows 操作系统，则可打开"记事本"编辑器；如果是其他操作系统，请在指导教师的帮助下打开一个纯文本编辑器。

步骤 2：按"程序模板"的要求输入源程序。

步骤 3：保存源文件，并命名为 Hello.java。将源文件保存到 C 盘的某个文件夹中，如 C:\1000。

步骤 4：打开命令行窗口来编译源文件。对于 Windows 操作系统，打开 MS-DOS 窗口；对于 Windows 2000/XP 操作系统，可以通过选择"开始"→"程序"→"附件"→"MS-DOS"来打开命令行窗口，也可以选择"开始"→"运行"，在弹出对话框的命令框中输入"cmd"命令来打开命令行窗口。如果目前 MS-DOS 窗口显示的逻辑符是"D:\"，输入"C:"并回车确认，使得当前 MS-DOS 窗口的状态是"C:\"。如果目前 MS-DOS 窗口的状态是 C 盘的某个子目录，请输入"cd\"，使得 MS-DOS 窗口的状态是"C:\"。当 MS-DOS 窗口的状态是"C:\"时，输入进入文件夹目录的命令，如"CD 1000"，然后执行下列编译命令：

C:\1000>\javac Hello.java

初学者在这一步可能遇到下列错误提示：

- Command not Found ——出现该错误的原因是没有设置好系统变量 Path，可参见主教材的 1.5 节内容。
- File not Found ——出现该错误的原因是没有将源文件保存在当前目录中（如 C:\1000），或源文件的名字不符合有关规定（如错误地将源文件命名为"hello.java"或"Hello.java.txt"）。注意：Java 语言的标识符是区分大小写的。
- 出现一些语法错误提示，如在中文输入状态下输入了程序中需要的分号等。Java 源程序中，语句所涉及的圆括号及标点符号都是英文状态下输入的，如"你好,很高兴学习 Java"中的引号必须是英文状态下的引号，而字符串里面的符号不受限制。

步骤 5：运行程序。

C:\1000> java Hello

初学者在本步骤可能遇到下列错误提示：Exception in thread "main" java.lang.NoClassFoundError。出现该错误的原因是没有设置好系统变量 Classpath（可参见主教材的 1.5 节内容），或者运行的不是主类的名字或程序没有主类。

6. 填写实验报告

实验报告的格式如下（可要求学生填写并由实验指导教师签字）：

学号：_____ 班级：_____ 姓名：_____ 时间：_____

实验内容	回答	教师评语
编译器怎样提示丢失花括号的错误		
编译器怎样提示语句丢失分号的错误		
如果在中文输入法状态下输入语句分号，程序编译时将出现错误，编译器怎样提示这一错误		
编译器怎样提示将 System 写成 system 这一错误		
编译器怎样提示将 String 写成 string 这一错误		

1.2.2 源文件的命名规则

1. 实验目的

本实验的目的是让学生掌握源文件的命名规则。

2. 实验要求

编写 Java 应用程序,其中有两个类:People 类和 A 类。A 类是主类,People 类是 public 类。

3. 程序效果示例

程序效果如图 1-2 所示。

4. 程序模板

按模板要求,将【代码】替换为程序代码。

图 1-2 运行主类

People.java

```
public class People{
    int height;
    void speak() {
        System.out.printf("我身高是: %d", height);
    }
}
class A{
    public static void main(String args[]) {
        People zhubajie;
        zhubajie=new People();
        zhubajie.height=170;
        【代码1】              //命令行窗口中输出 "zhubajie.height"
        【代码2】              //命令行窗口中输出 "主类不一定是 public 类"
        zhubajie.speak();
    }
}
```

5. 实验指导与检查

如果源文件中有多个类,那么只能有一个类是 public 类。如果有一个类是 public 类,那么源文件的名字必须与这个类的名字完全相同,扩展名是 .java(不要求主类必须是 public 类)。如果源文件没有 public 类,那么源文件的名字只要与某个类的名字相同,并且扩展名是 .java 即可。

Java 应用程序必须通过 Java 虚拟机中的 Java 解释器(java.exe)来解释执行其字节码文件。Java 应用程序总是从主类的 main()方法开始执行,因此必须按如下命令运行实验中的 Java 程序:

    C:\1000>java A

6. 填写实验报告

实验报告的格式如下(可要求学生填写并由实验指导教师签字):

学号：_____ 班级：_____ 姓名：_____ 时间：_____

实 验 内 容	回 答	教 师 评 语
将源文件保存为 A.java，编译器提示怎样的错误		
运行 People 类，解释器 java 提示怎样的错误		

1.3 知识扩展——联合编译

Java 程序的基本结构就是类，可以事先单独编译一个应用程序所需的类，将这些类和应用程序的主类存放在同一目录中即可。如果主类与其他类在同一目录中，则只需编译应用程序的主类。例如，有若干个源文件：Hello.java、A.java 和 B.java。每个源文件只有一个类，其中 Hello.java 是应用程序的主类（含有 main()方法），主类使用了类 A、B 和 C，那么只需编译源文件 Hello.java 即可。在编译 Hello.java 的过程中，Java 系统会自动先编译 A.java、B.java 和 C.java。

将下列 4 个源文件保存到同一目录中（如 C:\1000），然后编译 Hello.java。编译通过后，C:\1000 目录中将有 Hello.class、A.class 和 B.class 三个字节码文件。然后运行主类 Hello 即可。

Hello.java
```
    public class Hello{
        public static void main (String args[]){
            System.out.println("你好，很高兴学习 Java");
            A a=new A();
            a.fA();
            B b=new B();
            b.fB();
        }
    }
```
A.java
```
    public class A{
        void fA(){
            System.out.println("I am A");
        }
    }
```
B.java
```
    public class B{
        void fB(){
            System.out.println("I am B");
        }
    }
```

第 2 章

基本数据类型和数组

本章导读
- 知识概述
- 实验 1　输出俄文字母表
- 实验 2　从键盘输入数据
- 知识扩展——数组的快速复制

2.1 知识概述

1. 基本数据类型

Java 的基本数据类型包括：byte，short，int，long，float，double 和 char。要特别掌握基本类型的数据转换规则，基本数据类型按精度级别由低到高的顺序是：byte→short→int→long→float→double。

当把级别低的类型变量的值赋给级别高的类型变量时，系统自动完成数据类型的转换。当把级别高的类型变量的值赋给级别低的类型变量时，必须使用显式类型转换。

要观察一个字符在 Unicode 表中的顺序位置，必须使用 int 类型显式转换，如 (int)'a'。不可以使用 short 类型转换，因为 char 的最高位不是符号位。同样，要得到一个 0~65535 之间的数所代表的 Unicode 表中相应位置上的字符也必须使用 char 类型显式转换。char 类型数据与 byte、short、int 或 long 类型数据进行运算后的结果总是 int 类型数据。

2. 数组

数组属于引用类型数据，是将相同类型的数据按顺序组成的一种复合数据类型。可以用数组名加数组下标的方式来调用数组中的数据，下标从 0 开始。

2.2 实验练习

2.2.1 输出俄文字母表

1. 实验目的

本实验的目的是让学生掌握 char 类型数据与 int 类型数据之间的互相转换，同时了解 Unicode 字符表。

2. 实验要求

编写一个 Java 应用程序，该程序在命令行窗口中输出俄文字母表。

3. 程序效果示例

程序效果如图 2-1 所示。

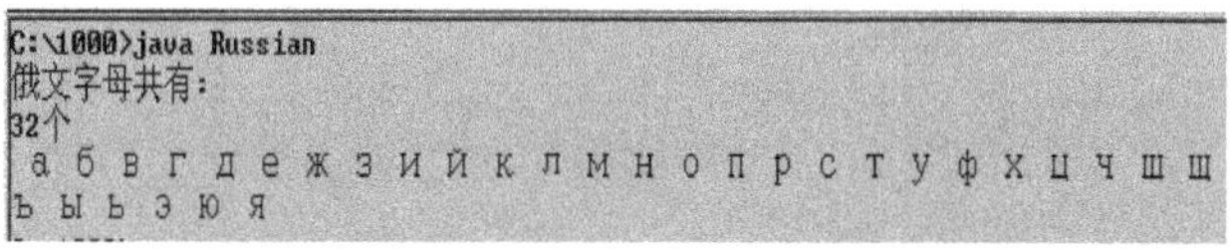

图 2-1 输出俄文字母

4. 程序模板

按模板要求，将【代码】替换为程序代码。

Russian.java

```
public class Russian{
    public static void main (String args[]) {
        int startPosition=0,endPosition=0;
```

```
        char cStart='α',cEnd='я';
【代码 1】            //对 cStart 进行 int 类型转换运算,并将结果赋值给 startPosition
【代码 2】            //对 cEnd 进行 int 类型转换运算,并将结果赋值给 endPosition
        System.out.println("俄文字母共有: ");
        System.out.println(endPosition-startPosition+ 1+ "个");
        for(int i=startPosition;i<=endPosition;i++){
            char c='\0';
【代码 3】            //对 i 进行 char 类型转换运算,并将结果赋值给 c
            System.out.print(" "+ c);
        }
    }
}
```

5. 实验指导与检查

⊙ 为了输出俄文字母表,首先获取俄文字母表的第一个字母和最后一个字母在 Unicode 表中的位置,然后使用循环输出其余俄文字母。

⊙ 向实验指导教师演示程序的运行效果。

6. 实验报告

实验报告的格式如下(可要求学生填写并由实验指导教师签字):

学号:_____ 班级:_____ 姓名:_____ 时间:_____

实验内容	回答	教师评语
将一个 float 类型数据直接赋值给 int 类型变量,程序编译时提示怎样的错误		
语句 　　byte x=128; 能编译通过吗?		
int　x=(byte)128; 程序输出变量 x 的值是多少		

2.2.2 从键盘输入数据

1. 实验目的

本实验的目的是让学生掌握从键盘输入基本类型的数据的方法。

2. 实验要求

编写一个 Java 应用程序,在主类的 main()方法中声明用于存放产品数量的 int 类型变量 amount 和产品单价的 float 类型变量,以及存放全部产品总价值的 float 类型变量 sum。

使用 Scanner 对象调用方法,让用户从键盘输入变量 amount、price 的值,然后计算出全部产品总价值,并输出 amount、prince、sum 的值。

```
输入产品数量(回车确认):18
输入产品单价(回车确认):12.89
数量:18,单价:12.89,总价值:232.02
```

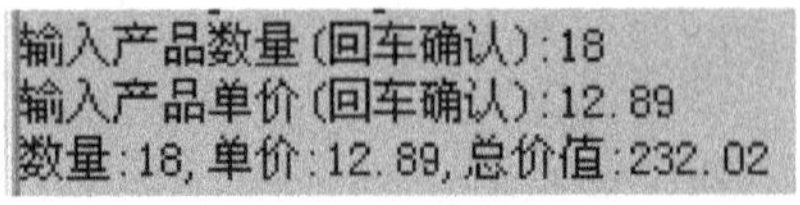

图 2-2 输入数量与价格

3. 程序效果示例

程序效果如图 2-2 所示。

4. 程序模板

按模板要求，将【代码】替换为程序代码。

InputData.java

```
import java.util.Scanner;
public class InputData{
    public static void main(String args[]){
        Scanner reader=new Scanner(System.in);
        int amount=0;
        float price=0,sum=0;
        System.out.print("输入产品数量(回车确认): ");
        【代码1】                    //从键盘输入 amount 的值
        System.out.print("输入产品单价(回车确认): ");
        【代码2】                    //从键盘输入 price 的值
        sum = price*amount;
        System.out.printf("数量: %d, 单价: %5.2f, 总价值: %5.2f",amount,price,sum);
    }
}
```

5. 实验指导与检查

- Scanner 对象调用 nextDouble()或 nextFloat()方法可以获取用户从键盘输入的浮点数。
- 向实验指导教师演示程序的运行效果。

6. 实验报告

实验报告的格式如下（可要求学生填写并由实验指导教师签字）：

学号：_____ 班级：_____ 姓名：_____ 时间：_____

实 验 内 容	回　答	教师评语
程序运行时，用户从键盘输入 abc，程序提示怎样的错误		

2.3 知识扩展——数组的快速复制

1. 数组的快速复制

我们已经知道，数组属于引用类型。也就是说，如果两个数组具有相同的引用，那么它们有完全相同的内存单元。例如：

　　int a[]={1,2}, b[];

如果执行

　　b=a;

那么 a 和 b 的值相同，即 a 的引用与 b 相同。这样，a[0]和 b[0]是相同的内存空间，a[1]和 b[1]的内存空间也相同。

有时我们想得到一个数组的"复制品"，即这个"复制品"数组与原数组的单元的个数相同，其中存储的数据也相同，但这个"复制品"数组单元值的改变不会影响到原数组，反之也是如此。

让 System 类调用类方法

  public static void arraycopy(sourceArray,int index1,copyArray,int index2,int length)

可以将数组 sourceArray 从索引 index1 开始后的 length 个单元中的数据复制到数组 copyArray 中，即将数组 sourceArray 中索引值从 index1 到 index1+length–1 单元中的数据复制到数组 copyArray 的某些单元中；copyArray 数组从第 index2 单元开始存放这些数据。如果数组 copyArray 不能存放下复制的数据，程序运行将发生异常。

  下面的 CopyArray.java 演示了 arraycopy()方法。

CopyArray.java

```java
class CopyArray {
    public static void main(String args[]){
        char a[]={'a', 'b', 'c', 'd', 'e', 'f'}, b[]={'1', '2', '3', '4', '5', '6'};
        int c[]={1, 2, 3, 4, 5, 6}, d[]={-1, -2, -3, -4, -5, -6};
        System.arraycopy(a, 0, b, 0, a.length);
        System.arraycopy(c, 2, d, 2, c.length-3);
        System.out.printf("\narray a:  ");
        for(int i=0;i<a.length;i++){
            System.out.printf("%3c", a[i]);
        }
        System.out.printf("\narray b:  ");
        for(int i=0;i<b.length;i++){
            System.out.printf("%3c", b[i]);
        }
        System.out.printf("\narray c:  ");
        for(int i=0;i<a.length;i++){
            System.out.printf("%3d", c[i]);
        }
        System.out.printf("\narray d:  ");
        for(int i=0;i<b.length;i++){
            System.out.printf("%3d", d[i]);
        }
    }
}
```

2．多维数组

  Java 采用"数组的数组"定义多维数组，一个二维数组由若干个一维数组组成。例如，二维数组

    int a[][]=new int [3][4];

就是由 3 个长度为 4 的一维数组构成的。

  构成二维数组的一维数组不必有相同的长度，在创建二维数组时可以分别指定构成该二维数组的一维数组的长度。例如：

    int a[][]=new int[3][];

二维数组 a 由 3 个一维数组 a[0]、a[1]和 a[2]构成。但它们的长度还没有确定，即这些一维数

组还没有分配内存空间，所以二维数组 a 还不能使用，必须创建这 3 个一维数组，如
  a[0]=new int[6];
  a[1]=new int[12];
  a[2]=new int[8];
也可直接用若干个一维数组初始化一个二维数组，这些一维数组的长度可以不尽相同。例如：
  int a[][]={{1},
    {1,1},
    {1,2,1},
    {1,3,3,1},
    {1,4,6,4,1},
    };

下面的 Example.java 应用程序输出杨辉三角形的前 5 行。

Example.java

```java
public class Example{
    public static void main(String args[]){
        int a[][]={{1},
                {1,1},
                {1,2,1},
                {1,3,3,1},
                {1,4,6,4,1},
                };
        for(int i=0;i<5;i++){
            for(int j=0;j<a[i].length;j++){
                System.out.printf("%4d",a[i][j]);
            }
            System.out.printf("%n");
        }
    }
}
```

第 3 章

运算符、表达式和语句

本章导读

- ✪ 知识概述
- ✪ 实验1 猜数字
- ✪ 实验2 回文数
- ✪ 知识扩展——使用 Arrays 类实现数组排序

3.1 知识概述

本章要求掌握各种运算符的使用规则，如算术运算符、关系运算符、布尔逻辑运算符、位运算符、赋值运算符等，掌握 Java 的表达式（特别要注意的是，一个 Java 表达式必须能求值），熟练使用 Java 的控制语句：条件分支语句和循环语句。

3.2 实验练习

3.2.1 计算电费

1．实验目的

本实验的目的是让学生使用 if-else 分支语句解决问题。

2．实验要求

为了节约用电，将用户的用电量分成 3 个区间，针对不同的区间给出不同的收费标准。对于 1～90 千瓦时（kW·h，度）的电量，每千瓦时 0.6 元；对于 91～150 千瓦时的电量，每千瓦时 1.1 元；对于大于 151 千瓦时的电量，每千瓦时 1.7 元。编写一个 Java 应用程序程序，在主类的 main() 方法中输入用户的用电量，程序输出电费。

3．程序效果示例

程序效果如图 3-1 所示。

```
输入电量:128
电费:95.80
```

图 3-1　计算电费

4．程序模板

按模板要求，将【代码】替换为 Java 程序代码。

Computer.java

```
import java.util.Scanner;
public class Computer{
    public static void main(String args[]) {
        Scanner reader=new Scanner(System.in);
        double amount=0;
        double price=0;
        System.out.print("输入电量：");
        amount=reader.nextDouble();
        if(amount<=90 && amount>=1){
            【代码1】              //计算 price 的值
        }
        else if(amount<=150 && amount>=91){
            【代码2】              //计算 price 的值
        }
        else if(amount>150){
            【代码3】              //计算 price 的值
        }
```

```
        else {
            System.out.println("输入电量: "+ amount+ "不合理");
        }
        System.out.printf("电费: %5.2f",price);
    }
}
```

5．实验指导与检查

- 要表达一个变量 x 的值是否在某个范围的时候，如小于–1 且大于–5 时，不要使用表达式 –5<x<–1，因为当 x 的值是–3 时，表达式–5<x<–1 的值是 false，应当使用表达式–5<x && x<–1 或 x>–5 && –1>x，显然当 x 的值是–3 时，这两个表达式的值都是 true。
- 向实验指导教师演示程序的运行效果。

6．实验报告

实验报告的格式如下（可要求学生填写并由实验指导教师签字）：

学号：_____ 班级：_____ 姓名：_____ 时间：_____

实验内容	回　　答	教师评语
在实验中省略 if-else if-else 语句中的 else 部分，是否有编译错误		
在实验中省略 if-else if-else 语句中的 else 部分，程序运行时用户输入–12，程序输出的结果是多少		

3.2.2　猜数字

1．实验目的

本实验的目的是让学生使用 if-else 分支语句和 while 循环语句解决问题。

2．实验要求

编写一个 Java 应用程序，实现如下功能：

（1）随机分配给客户一个 1～100 之间的整数。
（2）用户从键盘输入自己的猜测。
（3）程序返回提示信息，提示信息分别是"猜大了"、"猜小了"或"猜对了"。
（4）用户可根据提示信息再次输入猜测，直到提示信息是"猜对了"为止。

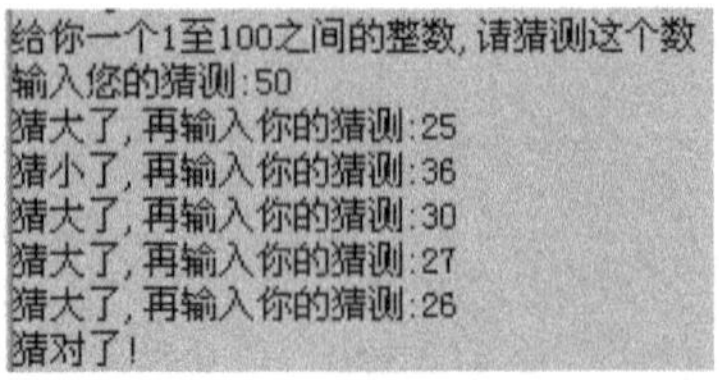

图 3-2　猜数字

3．程序效果示例

程序效果如图 3-2 所示。

4．程序模板

按模板要求，将【代码】替换为 Java 程序代码。

GuessNumber.java

```
import java.util.Scanner;
import java.util.Random;
public class GuessNumber{
    public static void main (String args[]){
```

```
Scanner reader=new Scanner(System.in);
Random random=new Random();
System.out.println("给你一个 1 至 100 之间的整数,请猜测这个数");
int realNumber=random.nextInt(100)+ 1;
                                    // random.nextInt(100)是[0,100)中的随机整数
int yourGuess = 0;
System.out.print("输入您的猜测: ");
yourGuess=reader.nextInt();
while(【代码1】){                          //循环条件
    if(【代码2】){                         //猜大了的条件代码
        System.out.print("猜大了,再输入你的猜测: ");
        yourGuess = reader.nextInt();
    }
    else if(【代码3】){                    //猜小了的条件代码
        System.out.print("猜小了,再输入你的猜测: ");
        yourGuess = reader.nextInt();
    }
}
System.out.println("猜对了! ");
    }
}
```

5. 实验指导与检查

- 我们经常使用一个 while 循环来"强迫"程序重复执行一段代码,【代码1】必须是求值为 boolean 类型数据的表达式。【代码1】的值为 true 时,让用户继续输入猜测;其值为 false 时,表明用户已经猜对了,就让用户停止输入猜测。
- 向实验指导教师演示程序的运行效果。

6. 实验报告

实验报告的格式如下(可要求学生填写并由实验指导教师签字):

学号:_____ 班级:_____ 姓名:_____ 时间:_____

实 验 内 容	回　　答	教师评语
用 "yourGuess>realNumber" 替换【代码1】,可以吗		
省略 if-else 语句中的 "yourGuess = reader.nextInt();",可以吗		
语句"System.out.println("猜对了!");"为何要放在 while 循环语句之后？放在 while 语句的循环体中合理吗		

3.3　知识扩展——使用 Arrays 类实现数组排序

可以使用循环实现对数组的排序,也可以使用循环查找一个数据是否在一个排序的数组中。这里省略数组的排序算法,让 Arrays 类调用类方法(有关类的知识见主教材第 4 章)来实现对数组的快速排序。

- 使用 java.util 包中的 Arrays 类的类方法 public static void sort(double a[]),可以把参数 a 指

定的 double 类型数组按升序排序。
- 使用 java.util 包中的 Arrays 类的静态方法 public static void sort(double a[],int start,int end)，可以把参数 a 指定的 double 类型数组中从位置 start 到 end 的值按升序排序。

在下面的 SortFind.java 中，首先将一个数组排序，然后使用折半法判断一个数是否在这个数组中。

SortFind.java

```java
import java.util.*;
public class SortFind{
    public static void main(String args[]){
        int n=0,start,end,middle;
        System.out.println("从键盘输入一个整数，程序将判断该数是否在一个数组中");
        int a[]={12,34,9,-23,45,6,45,90,123,19,34};
        Arrays.sort(a);
        Scanner reader=new Scanner(System.in);
        while(reader.hasNextInt()){                    //用户是否输入了整数
            n=reader.nextInt();
            start=0;
            end=a.length;
            middle=(start+end)/2;
            int count=0;
            while(n!=a[middle]){
                if(n>a[middle]){
                    start=middle;
                }
                else if(n<a[middle]){
                    end=middle;
                }
                middle=(start+end)/2;
                count++;
                if(count>a.length/2){
                    break;
                }
            }
            if(count>a.length/2){
                System.out.println(": "+n+"不在数组中");
            }
            else{
                System.out.println(": "+n+"是数组中的元素");
            }
            System.out.printf("%n可继续输入整数，或输入非整数结束程序%n");
        }
        System.out.println("你输入的数据不是整数");
    }
}
```

第 4 章　类 和 对 象

本章导读

- 知识概述
- 实验 1　机动车的类封装
- 实验 2　有理数的类封装
- 实验 3　公司和职员
- 实验 4　实例成员和类成员
- 实验 5　package 语句和 import 语句
- 知识扩展——Class 类的使用

4.1 知识概述

面向对象编程的核心思想之一就是将数据和对数据的操作封装在一起。抽象是指从具体的实例中抽取共同的性质形成某种一般的概念，如类的概念。人们经常谈到的机动车类就是从具体的实例中抽取共同的属性和功能形成的一个概念，而一个具体的轿车是机动车类的一个实例，即对象。对象将数据和对数据的操作合理、有效地封装在一起，如每辆轿车调用"加大油门"改变的都是自己的运行速度。本章要求掌握类的概念。类是组成 Java 程序的基本要素，类有两个重要的成员：成员变量和方法。类是创建对象的模板，类将对象的属性和功能封装为一个整体。

本章要求掌握实例成员变量和类变量的区别，不同对象的实例变量互不相同，而所有对象的类变量是相同的。

本章要求掌握实例方法与类方法的区别，实例方法必须由对象调用，而类方法既可以由对象调用，也可以由类来调用。

本章要求掌握方法重载的概念。方法重载是指一个类中可以有多个方法具有相同的名字，但这些方法的参数必须不同，即或者参数的个数不同，或者参数的类型不同。

本章要求掌握 import 语句和 package 语句。

本章要求了解成员变量和方法的访问权限。

4.2 实验练习

4.2.1 机动车的类封装

1．实验目的

本实验的目的是让学生学习使用类来封装对象的属性和功能。

2．实验要求

编写一个 Java 应用程序，该程序中有两个类：Vehicle（用于刻画机动车）和 User（主类）。具体要求如下：

（1）Vehicle 类有一个 double 类型的变量 speed（用于刻画机动车的速度）、一个 int 类型变量 power（用于刻画机动车的功率）；定义了 speedUp(int s)方法，体现机动车有加速功能；定义了 speedDown()等方法，体现机动车有减速功能；定义了 setPower(int p)方法，用于设置机动车的功率；定义了 getPower()方法，用于获取机动车的功率。

（2）在主类 User 的 main()方法中用 Vehicle 类创建对象，并让该对象调用想要的方法设置功率，演示加速和减速功能。

3．程序效果示例

程序效果如图 4-1 所示。

4．程序模板

按模板要求，将【代码】替换为程序代码。

Vehicle.java
```
public class Vehicle{
    【代码1】           //声明 double 类型变量 speed，刻画速度
    【代码2】           //声明 int 类型变量 power，刻画功率
    void speedUp(int s){
        【代码3】
        //将参数 s 的值与成员变量 speed 的和赋给成员变量 speed
    }
    void speedDown(int d){
        【代码4】           //将成员变量 speed 与参数 d 的差赋给成员变量 speed
    }
    void setPower(int p){
        【代码5】           //将参数 p 的值赋给成员变量 power
    }
    int getPower(){
        【代码6】           //返回成员变量 power 的值
    }
    double getSpeed(){
        return speed;
    }
}
```

```
car1的功率是：128
car2的功率是：76
car1目前的速度：80.0
car2目前的速度：100.0
car1目前的速度：70.0
car2目前的速度：80.0
```

图 4-1　Vehicle 类创建对

User.java
```
public class User{
    public static void main(String args[]){
        Vehicle car1,car2;
        【代码7】           //使用 new 运算符和默认的构造方法创建对象 car1
        【代码8】           //使用 new 运算符和默认的构造方法创建对象 car2
        car1.setPower(128);
        car2.setPower(76);
        System.out.println("car1 的功率是: "+car1.getPower());
        System.out.println("car2 的功率是: "+car2.getPower());
        【代码9】           //car1 调用 speedUp()方法将自己的 speed 的值增大 80
        【代码10】          //car2 调用 speedUp()方法将自己的 speed 的值增大 80
        System.out.println("car1 目前的速度: "+car1.getSpeed());
        System.out.println("car2 目前的速度: "+car2.getSpeed());
        car1.speedDown(10);
        car2.speedDown(20);
        System.out.println("car1 目前的速度: "+car1.getSpeed());
        System.out.println("car2 目前的速度: "+car2.getSpeed());
    }
}
```

5．实验指导与检查

⊙ 创建一个对象时，成员变量被分配内存空间，这些内存空间称为该对象的实体或变量。而对象中存放着引用，以确保这些变量由该对象操作使用。

- 空对象不能使用，即不能让一个空对象去调用方法产生行为。假如程序中使用了空对象，程序在运行时会出现异常：NullPointerException。由于对象动态地分配实体，所以 Java 的编译器对空对象不做检查。因此，在编写程序时要避免使用空对象。
- 将实验中的两个 Java 文件保存在同一目录中，分别编译或只编译主类 User.java，然后运行主类即可。
- 向实验指导教师演示程序的运行效果。

6．实验报告

实验报告的格式如下（可要求学生填写并由实验指导教师签字）：

学号：_____ 班级：_____ 姓名：_____ 时间：_____

实 验 内 容	回　　答	教 师 评 语
改进 speedUP()方法，使得 Vehicle 类的对象在加速时其 speed 值不能超过 200；改进 speedDown()方法，使得 Vehicle 类的对象在减速时其 speed 值不能小于 0		
增加一个刹车方法 void brake()，Vehicle 类的对象调用它能将 speed 值变为 0		

4.2.2　有理数的类封装

1．实验目的

本实验的目的是让学生学习使用对象进行分数的四则运算。

2．实验要求

我们有时希望程序能对分数（分子、分母都是整数）进行四则运算，而且两个分数进行四则运算后的结果仍然是分数。分数也称为有理数，是我们很熟悉的一种数。本实验要求用类实现对有理数的封装。有理数有两个重要的成员：分子和分母，还有重要的四则运算。

编写一个 Java 应用程序，该程序中有一个 Rational（有理数）类，具体要求如下。

（1）Rational 类有两个 int 类型的成员变量：numerator（分子）和 denominator（分母）。

（2）提供 Rational add(Rational r)方法，即有理数调用该方法与参数指定的有理数进行加法运算，并返回一个 Rational 对象。

（3）提供 Rational sub(Rational r)方法，即有理数调用该方法与参数指定的有理数进行减法运算，并返回一个 Rational 对象。

（4）提供 Rational muti(Rational r)方法，即有理数调用该方法与参数指定的有理数进行乘法运算，并返回一个 Rational 对象。

（5）提供 Rational div(Rational r)方法，即有理数调用该方法与参数指定的有理数进行除法运算，并返回一个 Rational 对象。

```
1/5+3/2 = 17/10
1/5-3/2 = -13/10
1/5×3/2 = 3/10
1/5÷3/2 = 2/15
```

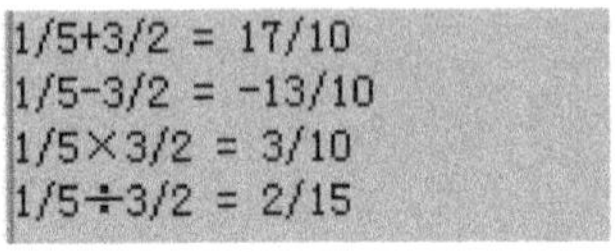

图 4-2　有理数的类封装

3．程序效果示例

程序效果如图 4-2 所示。

4．程序模板

按模板要求，将【代码】替换为程序代码。

Rational.java
```java
public class Rational{
    int numerator=1;                              //分子
    int denominator=1;                            //分母
    void setNumerator(int a){                     //设置分子
        int c=f(Math.abs(a),denominator);         //计算最大公约数
        numerator=a/c;
        denominator=denominator/c;
        if(numerator<0 && denominator<0){
            numerator=-numerator;
            denominator=-denominator;
        }
    }
    void setDenominator(int b){                   //设置分母
        int c=f(numerator,Math.abs(b));           //计算最大公约数
        numerator=numerator/c;
        denominator=b/c;
        if(numerator<0 && denominator<0){
            numerator=-numerator;
            denominator=-denominator;
        }
    }
    int getNumerator(){
        return numerator;
    }
    int getDenominator(){
        return denominator;
    }
    int f(int a,int b){                           //求 a 和 b 的最大公约数
        if(a==0){
            return 1;
        }
        if(a<b){
            int c=a;
            a=b;
            b=c;
        }
        int r=a%b;
        while(r!=0){
            a=b;
            b=r;
            r=a%b;
        }
        return b;
    }
```

```java
        Rational add(Rational r){                              //加法运算
            int a=r.getNumerator();                            //返回有理数 r 的分子
            int b=r.getDenominator();                          //返回有理数 r 的分母
            int newNumerator=numerator*b+ denominator*a;       //计算出新分子
            int newDenominator=denominator*b;                  //计算出新分母
            Rational result=new Rational();
            result.setNumerator(newNumerator);
            result.setDenominator(newDenominator);
            return result;
        }
        Rational sub(Rational r){                              //减法运算
            int a=r.getNumerator();
            int b=r.getDenominator();
            int newNumerator=numerator*b-denominator*a;
            int newDenominator=denominator*b;
            Rational result=new Rational();
            result.setNumerator(newNumerator);
            result.setDenominator(newDenominator);
            return result;
        }
        Rational muti(Rational r){                             //乘法运算
            int a=r.getNumerator();
            int b=r.getDenominator();
            int newNumerator=numerator*a;
            int newDenominator=denominator*b;
            Rational result=new Rational();
            result.setNumerator(newNumerator);
            result.setDenominator(newDenominator);
            return result;
        }
        Rational div(Rational r){                              //除法运算
            int a=r.getNumerator();
            int b=r.getDenominator();
            int newNumerator=numerator*b;
            int newDenominator=denominator*a;
            Rational result=new Rational();
            result.setNumerator(newNumerator);
            result.setDenominator(newDenominator);
            return result;
        }
    }
```

Computer.java

```java
    public class Computer{
        public static void main(String args[]){
            Rational r1=new Rational();
```

```
        【代码 1】                    //r1 设置分子是 1
        【代码 2】                    //r1 设置分母是 5
        Rational r2=new Rational();
        r2.setNumerator(3);
        r2.setDenominator(2);
        Rational result=【代码 3】      //r1 调用 add(Rational r)方法与 r2 进行加法运算
        int a=【代码 4】                //result 调用 getNumerator()方法返回分子
        int b=【代码 5】                //result 调用 getDenominator()方法返回分母
        System.out.println("1/5+ 3/2 = "+ a+ "/"+ b);
        result=r1.sub(r2);
        a=result.getNumerator();
        b=result.getDenominator();
        System.out.println("1/5-3/2="+ a+ "/"+ b);
        result=r1.muti(r2);
        a=result.getNumerator();
        b=result.getDenominator();
        System.out.println("1/5×3/2="+ a+ "/"+ b);
        result=r1.div(r2);
        a=result.getNumerator();
        b=result.getDenominator();
        System.out.println("1/5÷3/2="+ a+ "/"+ b);
    }
}
```

5. 实验指导与检查

- Math 类在 java.lang 包中，该类的 static 方法 abs(double x)可以返回 x 的绝对值。
- 将实验中的两个 Java 文件保存在同一目录中，分别编译或只编译主类 Computer.java，然后运行主类即可。
- 向实验指导教师演示程序的运行效果。

6. 实验报告

实验报告的格式如下（可要求学生填写并由实验指导教师签字）：

学号：_____ 班级：_____ 姓名：_____ 时间：_____

实 验 内 容	回 答	教师评语
在 Computer.java 中增加计算 1+3/2+5/3+8/5+13/8…前 20 项和的代码		

4.2.3 公司和职员

1. 实验目的

本实验的目的是让学生掌握对象的组合和参数传递。

2. 实验要求

编写一个 Java 应用程序，模拟公司与职员的关系，即公司将职员作为自己的成员。具体要求如下。

（1）有 3 个源文件：Employee.java、Corp.java 和 MainClass.java。其中，Employee.java 中的 Employee 类负责创建"职员"对象，Corp.java 中的 Corp 类负责创建"公司"对象，MainClass.java 是主类。

（2）在主类的 main()方法中首先使用 Employee 类创建一个对象 zhangLin，然后使用 Corp 类再创建一个对象 tomCorp，并将先前 Employee 类的实例 zhangLin 的引用传递给 tomCorp 对象的成员变量 secretary。

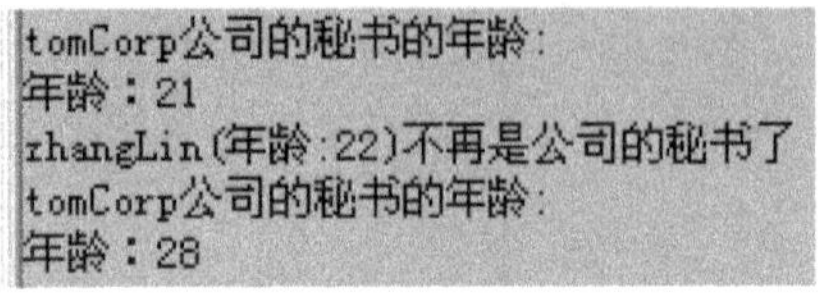

3. 程序效果示例

程序效果如图 4-3 所示。

图 4-3 公司和职员

4. 程序模板

按模板要求，将【代码】替换为程序代码。

Employee.java

```
public class Employee{
    int age;
    void setAge(int m){
        age=m;
    }
    void showAge(){
        System.out.println("年龄: "+age);
    }
}
```

Corp.java

```
public class Corp{
    Employee secretary;
    void setSecretary(Employee emp){
        【代码1】                              //将参数 emp 赋值给 secretary
    }
    void showSecretaryAge(){
        secretary.showAge();
    }
}
```

MainClass.java

```
public class MainClass{
    public static void main(String args[]){
        Employee zhangLin=new Employee();
        【代码2】     //zhangLin 调用 setAge(int m)，并向参数 m 传递整数 21
        Corp tomCorp = new Corp();
        【代码3】     //tomCorp 调用 setSecretary(Employee emp)，将 zhangLin 传递给参数
        tomCorp.setSecretary(zhangLin);
```

```
            System.out.println("tomCorp 公司的秘书的年龄: ");
            tomCorp.showSecretaryAge();
            zhangLin.setAge(22);
            Employee jiangHua=new Employee();
            jiangHua.setAge(28);
            tomCorp.setSecretary(jiangHua);
            System.out.printf("zhangLin(年龄: %d)不再是公司的秘书了\n",zhangLin.age);
            System.out.println("tomCorp 公司的秘书的年龄: ");
            tomCorp.showSecretaryAge();
      }
}
```

5. 实验指导与检查

- 类的成员变量可以是某个类的对象,如果用这样的类创建对象,那么该对象中就会有其他对象,也就是说,该类的对象将其他对象作为自己的组成部分。
- 当参数是引用类型时,"传值"传递的是变量中存放的"引用",而不是变量所引用的实体。注意,对于两个同类型的引用型变量,如果具有同样的引用,就会用同样的实体。因此,如果改变参数变量所引用的实体,就会导致原变量的实体发生同样的变化。
- 通过组合对象来复用方法也称为"黑盒"复用,因为当前对象只能委托所包含的对象调用其方法。这样,当前对象对所包含的对象的方法的细节是一无所知的。
- 向实验指导教师演示程序的运行效果。

6. 实验报告

实验报告的格式如下(可要求学生填写并由实验指导教师签字):

学号:_____ 班级:_____ 姓名:_____ 时间:_____

实 验 内 容	回　　答	教师评语
省略【代码2】,程序能否通过编译?若能通过编译,则程序输出的结果怎样		

4.2.4 实例成员和类成员

1. 实验目的

本实验的目的是让学生掌握类变量与实例变量、类方法与实例方法的区别。

2. 实验要求

按程序模板的要求编写源文件,特别注意程序的输出结果,并能正确地解释输出结果。

3. 程序效果示例

程序效果如图 4-4 所示。

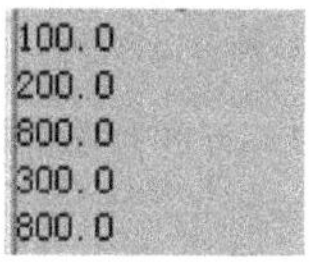

4. 程序模板

按模板要求,将【代码】替换为程序代码。

图 4-4　实例成员和类成员

Ex4_3.java

```
class A{
    【代码1】                        //声明一个float类型实例变量a
    【代码2】                        //声明一个float类型类变量(static变量)b
    void setA(float a){
        【代码3】                    //将参数a的值赋给成员变量a
    }
    void setB(float b){
        【代码4】                    //将参数b的值赋给成员变量b
    }
    float getA(){
        return a;
    }
    float getB(){
        return b;
    }
    void inputA(){
        System.out.println(a);
    }
    static void inputB(){
        System.out.println(b);
    }
}
public class Ex4_3{
    public static void main(String args[]){
        【代码5】                    //通过类名操作类变量b,并赋值100
        【代码6】                    //通过类名调用方法inputB()
        A cat=new A();
        A dog=new A();
        cat.setA(200);
        cat.setB(400);
        dog.setA(300);
        dog.setB(800);
        cat.inputA();
        cat.inputB();
        dog.inputA();
        dog.inputB();
    }
}
```

5．实验指导与检查

向实验指导教师演示程序的运行效果。

6．实验报告

实验报告的格式如下（可要求学生填写并由实验指导教师签字）：

学号： 班级： 姓名： 时间：		
实 验 内 容	回 答	教师评语
将 inputA()方法中的 　　System.out.println(a); 改写为 　　System.out.println(a+ b); 编译是否出错？为什么		
将 inputB()方法中的 　　System.out.println(b); 改写为 　　System.out.println(a+ b); 编译是否出错？为什么		

4.2.5 package 语句和 import 语句

1．实验目的

本实验的目的是让学生掌握 package 语句和 import 语句的方法。

2．实验要求

（1）按实验要求使用 package 语句，并通过 import 语句，使用 Java 平台提供的包中的类和自定义包中的类。

（2）掌握一些重要的操作步骤。

3．程序模板

（1）模板 1

将模板 1 给出的 Java 源文件命名为 Trangle.java，将编译后得到的字节码文件复制到目录 D:\shiyan\tom\jiafei 中。

Trangle.java

```java
package tom.jiafei;
public class Trangle{
    double sideA,sideB,sideC;
    boolean boo;
    public Trangle(double a,double b,double c){
        sideA=a;sideB=b;sideC=c;
        if(a+ b>c && a+ c>b && c+ b>a){
            System.out.println("我是一个三角形");
            boo=true;
        }
        else{
            System.out.println("我不是一个三角形");
            boo=false;
        }
    }
    public void 计算面积(){
```

```
            if(boo){
                double p=(sideA+ sideB+ sideC)/2.0;
                double area=Math.sqrt(p*(p-sideA)*(p-sideB)*(p-sideC)) ;
                System.out.println("是一个三角形，能计算面积");
                System.out.println("面积是： "+ area);
            }
            else{
                System.out.println("不是一个三角形，不能计算面积");
            }
        }
        public void 修改三边(double a, double b, double c){
            sideA=a;sideB=b;sideC=c;
            if(a+ b>c&&a+ c>b&&c+ b>a){
                boo=true;
            }
            else{
                boo=false;
            }
        }
    }
```

（2）模板 2

将模板 2 给出的 Java 源程序 SunRise.java 保存到目录 D:\shiyan 中。

SunRise.java

```
        import tom.jiafei.Trangle;
        import java.util.Date;
        class SunRise{
            public static void main(String args[]){
                Trangle trangle=new Trangle(12,3,104);
                trangle.计算面积();
                trangle.修改三边(3,4,5);
                trangle.计算面积();
                Date date=new Date();
                System.out.println(date);
            }
        }
```

4．实验指导与检查

⊙ 如果使用 import 语句引入了整个包中的类，那么可能增加编译时间，但绝对不会影响程序运行的性能。Java 运行平台由所需的 Java 类库和虚拟机组成，这些类库被包含在安装目录 jre\lib 的压缩文件 rt.jar 中。当程序执行时，Java 运行平台从类库中加载程序真正使用的类字节码到内存中。

⊙ 不可以将文件 Trangle.java 保存到目录 D:\shiyan 中。

⊙ 如果文件 Trangle.java 没有保存到目录 D:\shiyan\tom\jiafei 中，即不是保存在 SunRise 应用程序所在目录的 tom\jiafei 子目录中，就必须设置系统变量 Classpath 来指明包的位置。假设 Trangle.java 保存在目录 E:\hello\tom\jiafei 中，需要在命令行窗口中进行如下设置：
　　set classpath=E:\jdk1.6\jre\lib\rt.jar;.;E:\hello
⊙ 向实验指导教师演示程序的运行效果。

5. 实验报告

实验报告的格式如下（可要求学生填写并由实验指导教师签字）：

　　　　学号：_____　　班级：_____　　姓名：_____　　时间：_____

实 验 内 容	回　　答	教 师 评 语
如果不重新设置 classpath，编译 SunRise.java 时，系统提示怎样的错误		

4.3　知识扩展——Class 类的使用

1. 获取类的有关信息

Class 类封装对象运行时的状态。当类被加载创建对象时，与该类相关的一个类型为 Class 的对象就会自动创建，Class 类本身不提供构造方法，因此不能使用 new 运算符和构造方法显式地创建一个 Class 对象。任何对象调用 getClass()方法都可以获取与该对象相关的一个 Class 对象，这个 Class 对象调用如下方法可以获取创建对象的类的有关信息（如类的名字、类中的方法名称、成员变量的名称等）：

⊙ String getName() ——返回类的名字。
⊙ Constructor[] getDeclaredConstructors()——返回类的全部构造方法。
⊙ Field[] getDeclaredFields()——返回类的全部成员变量。
⊙ Method[] getDeclaredMethods()——返回类的全部方法。

下面的 ExampleOne.java 中有一个 A 类，我们使用相应的 Class 对象列出了 A 类的全部成员变量和方法的名称。

ExampleOne.java

```
    import java.lang.reflect.*;
    class A{
       int x;
       float y;
       double z;
       A(){
          x=12;
          y=12.901f;
          z=0.123456;
       }
       A(int x, float y, double z){
          this.x=x;
          this.y=y;
          this.z=z;
       }
```

```java
        public double getSum(){
            return x+ y+ z;
        }
        public void setx(int x){
            this.x=x;
        }
        public void setY(float y){
            this.y=y;
        }
        public void setZ(double z){
            this.z=z;
        }
    }
    public class ExampleOne {
        public static void main(String args[]){
            A a=new A(12, 34.9f, 0.54321);
            Class cs=a.getClass();
            String className=cs.getName();
            Constructor[] con=cs.getDeclaredConstructors();
            Field[] field=cs.getDeclaredFields() ;
            Method[] method=cs.getDeclaredMethods();
            System.out.println("类的名字: "+ className);
            System.out.println("类中有如下成员变量: ");
            for(int i=0;i<field.length;i++){
                System.out.println(field[i].toString());
            }
            System.out.println("类中有如下方法: ");
            for(int i=0;i<method.length;i++){
                System.out.println(method[i].toString() );
            }
            System.out.println("类中有如下构造方法: ");
            for(int i=0;i<con.length;i++){
                System.out.println(con[i].toString());
            }
        }
    }
```

2．使用 Class 实例化一个对象

即使没有使用类创建对象，我们也可以实例化一个与该类相关的 Class 对象，直接调用它的类方法

　　　　static Class forName(String className)

返回一个与参数 className 指定的类相关的 Class 对象。这个 Class 对象调用 newInstance()方法，就可以实例化一个 className 类的对象。注意：使用 Class 对象调用 newInstance()方法实例化一个 className 类的对象时，className 类必须有无参数的构造方法。

在下面的 ExampleTwo.java 中，分别使用 Class 对象来实例化一个 Rect 类和 Circle 类的对象。
ExampleTwo.java

```java
class Rect{
    private double width,height,area;
    public double getArea(){
        area=width*height;
        return area;
    }
    public void setWidth(double x){
        width=x;
    }
    public void setHeight(double y){
        height=y;
    }
}
class Circle{
    private double radius,area;
    public double getArea(){
        area=Math.PI*radius*radius;
        return area;
    }
    public void setRadius(double r){
        radius=r;
    }
}
public class ExampleTwo{
    public static void main(String args[]){
        try{
            Class cs=Class.forName("Rect");
            Rect rect=(Rect)cs.newInstance();
            rect.setWidth(100);
            rect.setHeight(10);
            System.out.println("rect 的面积"+ rect.getArea());
            cs=Class.forName("Circle");
            Circle circle=(Circle)cs.newInstance();
            circle.setRadius(100);
            System.out.println("circle 的面积"+ circle.getArea());
            cs=Class.forName("java.util.Date");
            java.util.Date date=(java.util.Date)cs.newInstance();
            System.out.println("现在时间： "+ date.toString() );
        }
        catch(Exception e){ }
    }
}
```

第 5 章

继承和接口

本章导读

✦ 知识概述
✦ 实验 1　继承
✦ 实验 2　上转型对象
✦ 实验 3　接口回调
✦ 实验 4　异常处理
✦ 知识扩展——可变参数和断言语句

5.1 知识概述

本章主要讲述了类的继承、多态、接口和内部类等重要概念。

子类继承父类的成员变量作为自己的一个成员变量，就好像它们是在子类中直接声明一样，可以被子类中自己声明的任何实例方法操作。也就是说，一个子类继承的成员应当是这个类的完全意义的成员，如果子类中声明的实例方法不能操作父类的某个成员变量，该成员变量就没有被子类继承。子类继承父类的方法作为子类中的一个方法，就像它们是在子类中直接声明一样，可以被子类中自己声明的任何实例方法调用。

多态是面向对象编程的又一重要特性。多态是指某个类的不同子类可以根据各自的需要重写父类的某个方法。这一部分内容涉及对象的上转型对象的知识。当把子类创建的对象的引用放到一个父类的对象中时，就得到了该对象的一个上转型对象，那么这个上转型对象在调用这个方法时就可能具有多种形态，因为不同的子类在重写父类的方法时可能产生不同的行为。注意，如果子类不能继承父类的某个方法，那么该方法不涉及重写问题。

接口是 Java 中非常重要的概念，必须很好地理解和掌握。接口的思想在于它可以增加类需要实现的功能，不同的类可以使用相同的接口，同一个类也可以实现多个接口。接口回调是多态的另一种体现。如果把使用某一接口的类创建的对象的引用赋给该接口声明的接口变量中，该接口变量就可以调用被类实现的接口中的方法，当接口变量调用被类实现的接口中的方法时，就是通知相应的对象调用接口的方法，这一过程称为对象功能的接口回调。不同的类在使用同一接口时，可能具有不同的功能体现，即接口的方法体不必相同，因此接口回调可能产生不同的行为。

内部类的使用可以使代码变得简洁方便。读者要特别掌握与类有关的匿名类以及与接口有关的匿名类。与类有关的匿名类就是该类的一个子类，与接口有关的匿名类就是实现了该接口的一个类。

5.2 实验练习

5.2.1 继承

1. 实验目的

本实验的目的是让学生巩固以下概念：

（1）子类的继承性。
（2）子类对象的创建过程。
（3）成员变量的继承和隐藏。
（4）方法的继承和重写。

2. 实验要求

编写程序模拟中国人、美国人、北京人。除主类外，程序中有 4 个类：People、ChinaPeople、AmericanPeople 和 BeijingPeople 类。要求如下：

（1）People 类包括权限是 protected 的 double 类型成员变量 height 和 weight，以及 public void speakHello()、public void averageHeight()和 public void averageWeight()方法。

（2）ChinaPeople 类是 People 的子类，新增 public void chinaGongfu()方法。要求 ChinaPeople 重写父类的 3 个方法：public void speakHello()、public void averageHeight()和 public void averageWeight()。

（3）AmericanPeople 类是 People 的子类，新增 public void americanBoxing()方法。要求 AmericanPeople 重写父类的 3 个方法：public void speakHello()、public void averageHeight()和 public void averageWeight()。

（4）BeijingPeople 类是 ChinaPeople 的子类，新增 public void beijingOpera()方法。要求 ChinaPeople 重写父类的 3 个方法：public void speakHello()、public void averageHeight()和 public void averageWeight()。

People、ChinaPeople、AmericanPeople 和 BeijingPeople 类的 UML 图如图 5-1 所示。

3．运行效果示例

运行效果如图 5-2 所示。

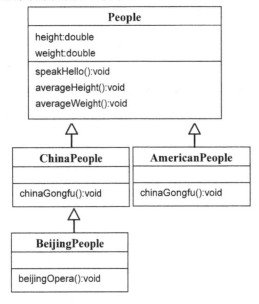

图 5-1　类的 UML 图　　　　图 5-2　成员的继承与重写

4．程序模板

按模板要求，将【代码】替换为 Java 程序代码。

People.java

```java
public class People{
    protected double weight,height;
    public void speakHello(){
        System.out.println("yayayaya");
    }
    public void averageHeight(){
        height=173;
        System.out.println("average height: "+ height);
    }
```

```
        public void averageWeight(){
            weight=70;
            System.out.println("average weight: "+ weight);
        }
    }
```

ChinaPeople.java
```
    public class ChinaPeople extends People{
        public void speakHello(){
            System.out.println("您好");
        }
        public void averageHeight(){
            height=168.78;
            System.out.println("中国人的平均身高: "+ height+ "厘米");
        }
        【代码1】   //重写 public void averageWeight()方法，输出"中国人的平均体重: 65 千克"
        public void chinaGongfu(){
            System.out.println("坐如钟，站如松，睡如弓");
        }
    }
```

AmericanPeople.java
```
    public class AmericanPeople extends People{
        【代码2】                  //重写 public void speakHello()方法，输出"How do you do"
        【代码3】
            //重写 public void averageHeight()方法，输出"American's average height: 176 cm"
        public void averageWeight(){
            weight=75;
            System.out.println("American's average weight: "+ weight+ "kg");
        }
        public void americanBoxing() {
            System.out.println("直拳、钩拳、组合拳");
        }
    }
```

BeijingPeople.java
```
    public class BeijingPeople extends ChinaPeople{
        【代码4】   //重写 public void averageHeight()方法，输出"北京人的平均身高: 172.5 厘米"
        【代码5】   //重写 public void averageWeight()方法，输出"北京人的平均体重: 70 千克"
        public void beijingOpera(){
            System.out.println("花脸、青衣、花旦和老生");
        }
    }
```

Example.java
```
    public class Example{
        public static void main(String args[]){
            ChinaPeople chinaPeople=new ChinaPeople();
```

```
            AmericanPeople americanPeople=new AmericanPeople();
            BeijingPeople beijingPeople=new BeijingPeople();
            chinaPeople.speakHello();
            americanPeople.speakHello();
            beijingPeople.speakHello();
            chinaPeople.averageHeight();
            americanPeople.averageHeight();
            beijingPeople.averageHeight();
            chinaPeople.averageWeight();
            americanPeople.averageWeight();
            beijingPeople.averageWeight();
            chinaPeople.chinaGongfu();
            americanPeople.americanBoxing();
            beijingPeople.beijingOpera();
            beijingPeople.chinaGongfu();
        }
    }
```

5. 实验指导与检查

- 如果子类可以继承父类的方法，子类就有权力重写这个方法。子类通过重写父类的方法可以改变方法的具体行为。
- 方法重写时一定要保证方法的名字、类型、参数的个数和类型同父类的某个方法完全相同，只有这样，子类继承的这个方法才被隐藏。
- 向实验指导教师演示程序的运行效果。

6. 实验报告

实验报告的格式如下（可要求学生填写并由实验指导教师签字）：

学号：_____ 班级：_____ 姓名：_____ 时间：_____

实 验 内 容	回 答	教师评语

就本程序而言，People 类中的
    public void speakHello()
    public void averageHeight()
    public void averageWeight()
这三个方法的方法体中的语句是否可以省略		

5.2.2 上转型对象

1. 实验目的

本实验的目的是让学生掌握上转型对象的使用。在讲述继承和多态时，主教材通过子类对象的上转型体现了继承的多态性，即把子类创建的对象的引用放到一个父类的对象中，得到该对象的一个上转型对象，那么这个上转型对象在调用方法时就可能具有多种形态。不同对象的上转型对象调用同一方法可能产生不同的行为。

2. 实验要求

编写一个抽象类 Employee，其子类有 YearWorker、MonthWorker 和 WeekWorker。YearWorker 对象按年领取工资，MonthWorker 按月领取工资，WeekWorker 按周领取工资。Employee 类有一个抽象方法 public abstract earnings()。

子类必须重写父类的 earnings() 方法，给出各自领取工资的具体方式。

编写一个类 Company，该类用 employee 数组作为成员。employee 数组的单元可以是 YearWorker 对象、MonthWorker 对象或 WeekWorker 对象的上转型对象。程序能输出 Company 对象一年需要支付的工资总额。

3. 运行效果示例

运行效果如图 5-3 所示。

图 5-3 上转型对象

4. 程序模板

按模板要求，将【代码】替换为程序代码。

HardWork.java

```
    abstract class Employee{
        public abstract double earnings();
    }
    class YearWorker extends Employee{
        【代码1】                          //重写 earnings()方法
    }
    class MonthWorker extends Employee{
        【代码2】                          //重写 earnings()方法
    }
    class WeekWorker extends Employee{
        【代码3】                          //重写 earnings()方法
    }
    class Company{
        Employee[] employee;
        double salaries=0;
        Company(Employee[] employee){
            this.employee=employee;
        }
        public double salariesPay(){
            salaries=0;
            【代码4】                      //计算 salaries
            return salaries;
        }
    }
    public class HardWork{
        public static void main(String args[]){
            Employee[] employee=new Employee[20];
            for(int i=0;i<employee.length;i++){
                if(i%3==0){
```

```
                employee[i]=new WeekWorker();
            }
            else if(i%3==1){
                employee[i]=new MonthWorker();
            }
            else if(i%3==2){
                employee[i]=new YearWorker();
            }
        }
        Company company=new Company(employee);
        System.out.println("公司年工资总额: "+company.salariesPay());
    }
}
```

5. 实验指导与检查

- 尽管抽象类不能创建对象，但抽象类声明的对象可以存放子类对象的引用，即成为子类对象的上转型对象。由于抽象类可以有抽象方法，这样就保证子类必须重写这些抽象方法。由于数组 employee 的每个单元都是某个子类对象的上转型对象，实验中的【代码4】可以通过循环语句，让数组 employee 的每个单元调用 earnings()方法，并将该方法返回的值累加到 salaries。
- 向实验指导教师演示程序的运行效果。

6. 实验报告

实验报告的格式如下（可要求学生填写并由实验指导教师签字）：

学号：_____ 班级：_____ 姓名：_____ 时间：_____

实 验 内 容	回　　答	教 师 评 语
子类 YearWorker 如果不重写 earnings()方法，程序编译时将提示怎样的错误		

5.2.3 接口回调

1. 实验目的

本实验的目的是让学生掌握接口回调技术。

2. 实验要求

编写一个接口 ComputeTotalSales，该接口中有方法 public double totalSalesByYear()，有 3 个实现该接口的类 Television，Computer 和 Mobile。这 3 个类通过实现接口 computeTotalSales，给出自己的年销售额。

编写一个类 Shop，该类用 computeTotalSales 数组作为成员，computeTotalSales 数组的单元可以存放 Television 对象的引用、Computer 对象的引用或 Mobile 对象的引用。程序能输出 Shop 对象的年销售额。

3. 运行效果示例

运行效果如图 5-4 所示。

4．程序模板

按模板要求，将【代码】替换为程序代码。

HappySale.java

```
interface ComputeTotalSales{
    public double totalSalesByYear();
}
class Television implements ComputeTotalSales{
    【代码1】                          //实现totalSalesByYear()方法
}
class Computer implements  ComputeTotalSales{
    【代码2】                          //实现totalSalesByYear()方法
}
class Mobile implements  ComputeTotalSales{
    【代码3】                          //实现totalSalesByYear()方法
}
class Shop {
    ComputeTotalSales[] goods;
    double totalSales=0;
    Shop(ComputeTotalSales[] goods){
        this.goods=goods;
    }
    public double giveTotalSales(){
        totalSales=0;
        【代码4】                       //计算totalSales
        return totalSales;
    }
}
public class HappySale{
    public static void main(String args[]){
        ComputeTotalSales[] goods=new ComputeTotalSales[50];
        for(int i=0;i<goods.length;i++){
            if(i%3==0){
                goods[i]=new Television();
            }
            else if(i%3==1){
                goods[i]=new Computer();
            }
            else if(i%3==2){
                goods[i]=new Mobile();
            }
        }
        Shop shop=new Shop(goods);
        System.out.println("商店年销售额： "+ shop.giveTotalSales());
    }
}
```

图 5-4 接口回调

5．实验指导与检查

⊙ 由于数组 goods 的每个单元存放的是实现 ComputeTotalSales 接口的对象的引用，程序中的【代码 4】通过循环语句，可以让数组 goods 的每个单元调用 totalSalesByYear()方法，并将该方法返回的值累加到 totalSales。

⊙ 向实验指导教师演示程序的运行效果。

6．实验报告

实验报告的格式如下（可要求学生填写并由实验指导教师签字）：

学号：_____ 班级：_____ 姓名：_____ 时间：_____

实　验　内　容	回　　答	教　师　评　语
类 Mobile 如果不实现 totalSalesByYear()方法，程序编译时提示怎样的错误		

5.2.4 异常处理

1．实验目的

本实验的目的是让学生学习怎样定义异常类和抛出异常。

2．实验要求

声明两个 Exception 的异常子类：NoLowerLetter 类和 NoDigit 类。再声明一个 People 类，该类中的 void printLetter(char c)方法抛出 NoLowerLetter 异常类对象，void printDigit(char c)方法抛出 NoDigit 异常类对象。

3．运行效果示例

运行效果如图 5-5 所示。

图 5-5　处理异常

4．程序模板

按模板要求，将【代码】替换为程序代码。

ExceptionExample.java

```
【代码 1】{                          //类声明，声明一个 Exception 的子类 NoLowerLetter
    public void print(){
        System.out.printf("%c",'#');
    }
}
【代码 2】{                          //类声明，声明一个 Exception 的子类 NoDigit
    public void print(){
        System.out.printf("%c",'*');
    }
}
class People{
    void printLetter(char c) throws NoLowerLetter{
        if(c<'a' || c>'z'){
            NoLowerLetter noLowerLetter=【代码 3】    //创建 NoLowerLetter 类型对象
```

```
                【代码 4】                           //抛出 noLowerLetter
            }
            else{
                System.out.print(c);
            }
        }
        void printDigit(char c) throws NoDigit{
            if(c<'1' || c>'9'){
                NoDigit noDigit=new NoDigit();
                throw noDigit
            }
            else{
                System.out.print(c);
            }
        }
    }
    public class ExceptionExample{
        public static void main (String args[]){
            People people=new People();
            for(int i=0;i<128;i++){
                try{
                    people.printLetter((char)i);
                }
                catch(NoLowerLetter e){
                    e.print();
                }
            }
            for(int i=0;i<128;i++){
                try{
                    people.printDigit((char)i);
                }
                catch(NoDigit e){
                    e.print();
                }
            }
        }
    }
```

5. **实验指导与检查**

- Java 运行环境定义了许多异常类（Exception 的子类），当对应的异常发生时，就用相应的异常类创建一个异常对象，并等待处理。
- 可以扩展 Exception 类定义自己的异常类，然后规定哪些方法产生这样的异常。一个方法在声明时可以使用 **throws** 关键字声明抛出所要产生的若干个异常，并在方法体中具体给出产生异常的操作，即用相应的异常类创建对象，该方法抛出所创建的异常对象来结束方法的执行。程序必须在 **try-catch** 块语句中调用抛出异常的方法。
- 向实验指导教师演示程序的运行效果。

6. **实验报告**

实验报告的格式如下（可要求学生填写并由实验指导教师签字）：

学号：_____ 班级：_____ 姓名：_____ 时间：_____

实验内容	回　　答	教师评语

下述代码输出的结果是什么
　try{
　　　for(int i=0;i<128;i++){
　　　　people.printLetter((char)i);
　　　}
　}
　catch(NoLowerLetter e){ e.print(); }		

5.3　知识扩展——可变参数和断言语句

1. 可变参数（variable argument）

SDK 1.5 新增了可变参数功能。可变参数是指：在声明方法时，不指定方法的参数的个数，代替地使用"..."。例如：

　　　public void f(int ... x)

那么，方法 f()的参数类型是 int 类型，但个数不确定。这里称 x 是一个"参数代表"。参数代表可以通过下标运算表示参数列表中的具体参数，如 x[0]和 x[1]分别代表参数列表中的第一个参数和第二个参数，x.length 等于参数的个数。参数代表非常类似我们自然语言中的"等"（英语中的 and so on），我们经常说"汽车、卡车等"。

对于参数类型相同的方法，使用可变参数可以使方法的调用更灵活。下面的 A 类的两个方法使用了可变参数。

Example.java

```java
class A{
    public void f(int ... x) {
        int sum=0;
        for(int i=0;i<x.length;i++) {
            sum=sum+ x[i];
        }
        System.out.println(sum);
    }
    public void g(String ...s) {
        for(int i=0;i<s.length;i++){
            System.out.printf("%s",s[i]);
        }
    }
}
public class Example {
    public static void main(String args[]) {
        A a=new A();
        a.f(1,2,3,4,5);
        a.f(-1, -2, -3, -4, -5, -6);
        a.g("how","are","you");
    }
}
```

2. 断言（assert）语句

断言语句是 Java 语言在 SDK 1.4 以后新增的一项功能。断言语句在调试代码阶段非常有用，在程序正式运行时可以关闭断言语句。

断言语句使用关键字 assert 来声明，断言语句有以下两种格式：

assert booleanExpression;

assert booleanExpression : messageException;

其中，booleanExpression 必须是求值为 boolean 类型的表达式，messageException 可以是求值为字符串的表达式。

如果使用

assert booleanExpression;

形式的断言语句，当 booleanExpression 的值是 false 时，程序从断言语句处停止执行；当 booleanExpression 的值是 true 时，程序从断言语句处继续执行。

如果使用

assert booleanExpression : messageException;

形式的断言语句，当 booleanExpression 的值是 false 时，程序从断言语句处停止执行，并输出 messageException 表达式的值，提示用户出现了怎样的问题；当 booleanExpression 的值是 true 时，程序从断言语句处继续执行。

断言语句一般用于程序不准备通过捕获异常来处理的错误，如当发生某个错误时，要求程序必须立即停止执行，在调试代码阶段让断言语句发挥作用，这样就可以发现一些致命的错误。当程序正式运行时就可以关闭断言语句，但仍把断言语句保留在源代码中，如果以后应用程序又需要调试，可以重新启用断言语句。

当使用 Java 解释器直接运行应用程序时，默认将关闭断言语句。在 SDK 1.4 中，可以通过 Java 命令的命令行选项"-ea"或"-enableassertion"来启用断言语句。例如：

java -ea mainClass

或

java -enableassertion mainClass

在调试程序时，可以使用"-ea"启用断言语句，使用"-da"显式地关闭断言语句。例如，通过 Java 命令的命令行选项"-da"或"-disableassertion"关闭断言语句：

java -da mainClass

或

java -disableassertion mainClass

下面的 AssertExample.java 源程序在计算平方根时使用了断言语句。

AssertExample.java

```
public class AssertExample {
    public static void main(String args[]) {
        double x=-6;
        double y=Math.sqrt(x);
        assert x>0:"负数不能计算平方根";
        System.out.println(y);
    }
}
```

第 6 章 字符串和正则表达式

本章导读

- ✿ 知识概述
- ✿ 实验 1　String 类的常用方法
- ✿ 实验 2　StringBuffer 类的常用方法
- ✿ 实验 3　Scanner 类与字符串分解
- ✿ 实验 4　模式匹配
- ✿ 知识扩展——元词和定位元字符

6.1 知识概述

本章主要讲述了 String 类、StringBuffer 类、StringTokenizer 类以及正则表达式和模式匹配。String 类和 StringBuffer 类封装处理字符串有关的操作，两者的区别是，String 对象的实体是不可改变的，而 StringBuffer 对象的实体可以发生变化。两者之间也存在着密切的联系，可以用 String 对象作为参数创建一个 StringBuffer 对象，如

   StringBuffer buffer=new StringBuffer("hello");

也可以用 StringBuffer 对象作为参数创建一个 Stringr 对象，如

   String str=new String(new StringBuffer("how are you"));

读者要很好地掌握 String 类和 Stringbuffer 类的常用方法。

StringTokenizer 类封装了分解字符串的简单方法，利用 StringTokenizer 对象可以分解出字符串中的语言符号。

Scanner 类的实例从字符串中解析数据。在默认情况下，Scanner 对象将空格作为分隔标记解析字符串。Scanner 对象调用方法 useDelimiter(String regex)，将正则表达式作为分隔标记，即 Scanner 对象在解析字符串时把与正则表达式 regex 匹配的字符串作为分隔标记。

模式匹配是重要的内容。一个含有特殊元字符的字符串 patternString 称为一个正则表达式。为了比较一个字符串 matchString 是否匹配于正则表达式，必须使用正则表达式 patternString 创建一个模式对象：

   Pattern p= Pattern.compile(patternString);

然后创建匹配对象：

   Matcher m;
   m=p.matcher(matchString);

那么，对象 m 就可以调用一些方法返回有关模式匹配的信息。例如，m 调用 find()方法可以寻找 matchString 和 patternString 匹配的下一子序列，如果成功，则该方法返回 true，否则返回 false。m 首次调用该方法时，寻找 matchString 中第一个与 patternString 匹配的子序列，如果 find()方法返回 true，m 再调用 find()方法，将从上一次匹配模式成功的子序列后寻找下一个匹配模式的子字符串。

6.2 实验练习

6.2.1 String 类的常用方法

1．实验目的

本实验的目的是让学生掌握 String 类的常用方法。

2．实验要求

编写一个 Java 应用程序，判断两个字符串是否相同，判断字符串的前缀、后缀是否与某个字符串相同，以及按字典序比较两个字符串的大小关系，进行字符串检索，创建子字符串，将数字型字符串转化为数字，将字符串存放到数组中，用字符数组创建字符串等。

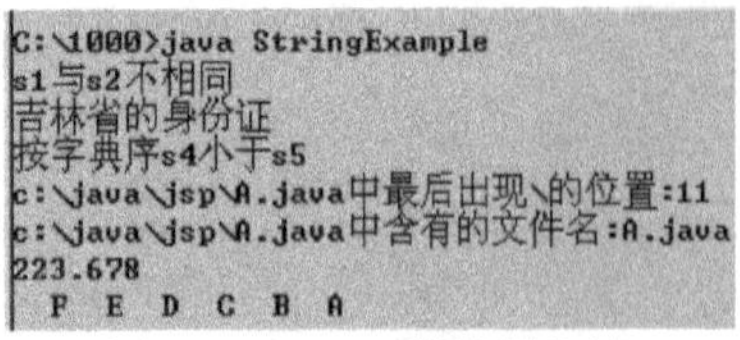

3. 运行效果示例

运行效果如图 6-1 所示。

4. 程序模板

图 6-1 String 类常用方法

按模板要求，将【代码】替换为程序代码。

StringExample.java

```java
class StringExample{
    public static void main(String args[]){
        String s1=new String("you are a student"), s2=new String("how are you");
        if( 【代码1】 ){                          //判断 s1 与 s2 是否相同
            System.out.println("s1 与 s2 相同");
        }
        else{
            System.out.println("s1 与 s2 不相同");
        }
        String s3=new String("220302198510220240");
        if( 【代码2】 ){                          //判断 s3 的前缀是否为"220302"
            System.out.println("吉林省的身份证");
        }
        String s4=new String("你"), s5=new String("我");
        if( 【代码3】 ){                          //按字典序，s4 大于 s5 的表达式
            System.out.println("按字典序，s4 大于 s5");
        }
        else{
            System.out.println("按字典序，s4 小于 s5");
        }
        int position=0;
        String path="C:\\java\\jsp\\A.java";
        position= 【代码4】                       //获取 path 中最后出现目录分隔符号的位置
        System.out.println("C:\\java\\jsp\\A.java 中最后出现\\的位置："+ position);
        String fileName= 【代码5】                //获取 path 中的"A.java"子字符串
        System.out.println("C:\\java\\jsp\\A.java 中含有的文件名："+ fileName);
        String s6=new String("100"), s7=new String("123.678");
        int n1= 【代码6】                         //将 s6 转化成 int 类型数据
        double n2= 【代码7】                      //将 s7 转化成 double 类型数据
        double n=n1+ n2;
        System.out.println(n);
        String s8=new String("ABCDEF");
        char a[]= 【代码8】                       //将 s8 存放到数组 a 中
        for(int i=a.length-1;i>=0;i--){
            System.out.printf("%3c",a[i]);
        }
    }
}
```

5. 实验指导与检查

- 字符串 s 调用方法 substring()返回一个新的字符串对象，而 s 本身不会发生变化。
- 向实验指导教师演示程序的运行效果。

6. 实验报告

实验报告的格式如下（可要求学生填写并由实验指导教师签字）：

学号：_____ 班级：_____ 姓名：_____ 时间：_____

实验内容	回答	教师评语
程序中的 s6 改写成 　　String s6=new String("1a12b"); 运行时提示怎样的错误		
请用数组 a 的前 3 个单元创建一个字符串并输出该串		
请给出获取 path 中 "jsp" 子字符串的代码		

在程序的适当位置增加如下代码，注意输出的结果
　　String str1=new String("ABCABC"),
　　　　　str2=null,
　　　　　str3=null,
　　　　　str4=null;
　　str2=str1.replaceAll("A", "First");
　　str3=str2.replaceAll("B", "Second");
　　str4=str3.replaceAll("C", "Third");
　　System.out.println(str1);
　　System.out.println(str2);
　　System.out.println(str3);
　　System.out.println(str4);		

6.2.2　StringBuffer 类的常用方法

1. 实验目的

本实验的目的是让学生掌握 StringBuffer 类的常用方法。

2. 实验要求

编写一个应用程序，使用 StringBuffer 对象实现对字符串的编辑操作，如替换字符串中的某些字符、删除字符串中的某些字符、在字符串中插入或末尾添加新的字符串等。

3. 运行效果示例

运行效果如图 6-2 所示。

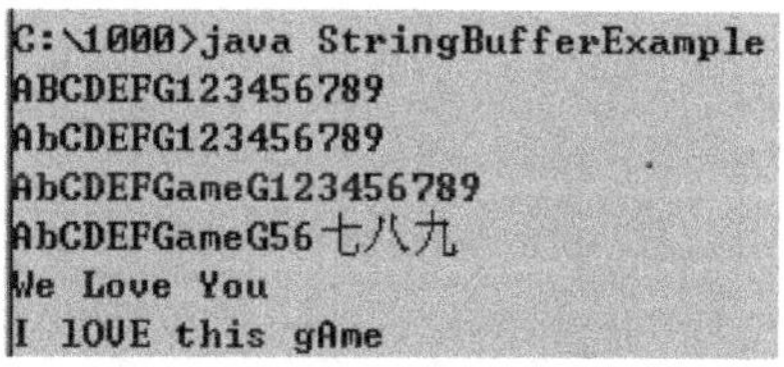

图 6-2　StringBuffer 类常用方法

4. 程序模板

按模板要求，将【代码】替换为程序代码。

StringBufferExample.java

```
class StringBufferExample{
    public static void main(String args[]){
        StringBuffer str=new StringBuffer("ABCDEFG");
        【代码1】                    //向 str 末尾添加"123456789"
```

```
                System.out.println(str);
                【代码2】                              //将 str 中的字符'B'替换为'b'
                System.out.println(str);
                【代码3】                              //在 str 中的"123456789"前面插入"Game"
                System.out.println(str);
                int index=【代码4】                    //获取 str 中首次出现'1'的位置
                【代码5】                              //删除 str 中的"1234"
                int n=【代码6】                        //获取 str 中的字符个数
                【代码7】                              //将 str 中的"789"替换为"七八九"
                System.out.println(str);
                StringBuffer otherStr=new StringBuffer("we love you");
                int start=0;
                char c='\0';
                while(start!=-1){
                    if(start!=0)
                        start=start+1;
                    c=otherStr.charAt(start);
                    if(Character.isLowerCase(c)){
                        c=Character.toUpperCase(c);
                        otherStr.setCharAt(start,c);
                    }
                    start=otherStr.indexOf(" ",start);          //查找下一个空格
                }
                System.out.println(otherStr);
                StringBuffer yourStr=new StringBuffer("i Love THIS GaME");
                for(int i=0;i<yourStr.length();i++){
                    char c1=yourStr.charAt(i);
                    if(Character.isLowerCase(c1)){
                        c1=Character.toUpperCase(c1);
                        yourStr.setCharAt(i,c1);
                    }
                    else if(Character.isUpperCase(c1)){
                        c1=Character.toLowerCase(c1);
                        yourStr.setCharAt(i,c1);
                    }
                }
                System.out.println(yourStr);
            }
        }
```

5. 实验指导与检查

- StringBuffer 字符串 s 调用方法 insert()会使得 s 的实体发生变化。
- 向指导教师演示程序的运行效果。

6. 实验报告

实验报告的格式如下（可要求学生填写并由实验指导教师签字）：

学号：_____ 班级：_____ 姓名：_____ 时间：_____		
实验内容	回 答	教师评语
StringBuffer 类的 length()方法与 capacity()方法有何不同		
上述程序的最后添加如下代码： 　System.out.println(str.length()); 　System.out.println(str.capacity()); 输出结果是什么		

6.2.3 Scanner 类与字符串分解

1．实验目的
本实验的目的是掌握怎样使用 Scanner 类的对象从字符串中解析程序所需要的数据。

2．实验要求
菜单的内容为："北京烤鸭：189 元　西芹炒肉：12.9 元　酸菜鱼：69 元　铁板牛柳：32 元"。编写一个 Java 应用程序，输出菜单中的价格数据，并计算出菜单的总价格。

3．运行效果示例
运行效果如图 6-3 所示。

4．程序模板
按模板要求，将【代码】替换为程序代码。

图 6-3 菜单的价格

ComputerPrice.java

```
import java.util.*;
public class ComputePrice{
    public static void main(String args[]){
        String menu="北京烤鸭：189元　西芹炒肉：12.9元　酸菜鱼：69元　铁板牛柳：32元";
        Scanner scanner=【代码1】
                          //Scanner 类创建 scanner，将 menu 传递给构造方法的参数
        String regex="[^0123456789.]+";
        【代码2】  //scanner 调用 useDelimiter(String regex)，将 regex 传递给该方法的参数
        double sum=0;
        while(scanner.hasNext()){
            try{
                double price=【代码3】      //scanner 调用 nextDouble()返回数字单词
                sum=sum+ price;
                System.out.println(price);
            }
            catch(InputMismatchException exp){
                String t=scanner.next();
            }
        }
        System.out.println("菜单总价格："+ sum+ "元");
    }
}
```

5. 实验指导与检查

- scanner 可以用 nextInt()方法或 nextDouble()方法解析字符串中的数字型的单词，即 scanner 可以调用 nextInt()方法或 nextDouble()方法，将数字型单词转化为 int 或 double 类型数据返回。如果单词不是数字型单词，scanner 调用 nextInt()方法或 nextDouble()方法，将发生 InputMismatchException 异常。
- 向指导教师演示程序的运行效果。

6. 实验报告

实验报告的格式如下（可要求学生填写并由实验指导教师签字）：

学号：_____ 班级：_____ 姓名：_____ 时间：_____

实 验 内 容	回 答	教 师 评 语
用"[^(http//|www)\56?\\w+\56{1}\\w+\56{1}\\p{Alpha}]+"作为分隔标记，解析"中央电视台 www.cctv.com、音乐网站 www.violin.com"中的全部网站链接地址		

6.2.4 模式匹配

1. 实验目的

本实验的目的是让学生掌握模式匹配的基本方法。

2. 实验要求

下列字符串中，将"登录网站"错写为"登陆网站"，将"惊慌失措"错写为"惊慌失错"，即原文为："忘记密码，不要惊慌失错，请登陆 www.yy.cn 或登陆 www.tt.cc"。编写一个 Java 应用程序，把错别字替换为正确的用字，并输出替换后的字符串。

3. 运行效果示例

运行效果如图 6-4 所示。

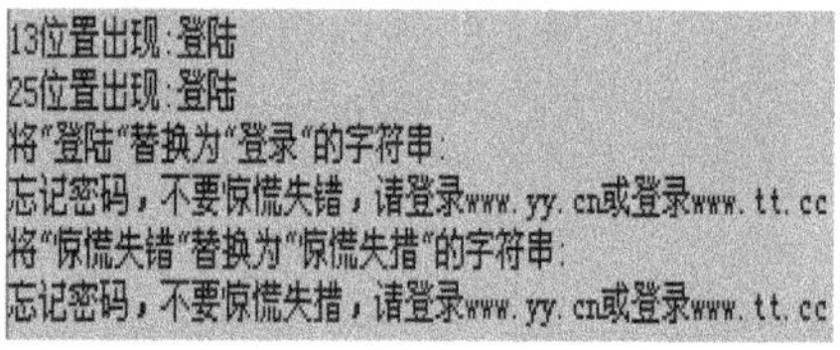

4. 程序模板

按模板要求，将【代码】替换为程序代码。

图 6-4 替换错别字

ReplaceErrorWord.java

```
import java.util.regex.*;
public class ReplaceErrorWord{
    public static void main(String args[]){
        String str="忘记密码，不要惊慌失错，请登陆 www.yy.cn 或登陆 www.tt.cc";
        Pattern pattern;
        Matcher matcher;
        String regex="登陆";
        pattern=【代码1】              //使用 regex 初试化模式对象 pattern
        matcher=【代码2】              //得到检索 str 的匹配对象 matcher
        while(matcher.find()){
            String s=matcher.group();
            System.out.print(matcher.start()+ "位置出现: ");
            System.out.println(s);
```

```
            }
            System.out.println("将\"登陆\"替换为\"登录\"的字符串：");
            String result=matcher.replaceAll("登录");
            System.out.println(result);
            pattern=Pattern.compile("惊慌失错");
            matcher=pattern.matcher(result);
            System.out.println("将\"惊慌失错\"替换为\"惊慌失措\"的字符串：");
            result=matcher.replaceAll("惊慌失措");
            System.out.println(result);
        }
    }
```

5. 实验指导与检查

⊙ matcher 调用 boolean matches()方法判断 str 是否完全与 regex 匹配。matcher 调用 boolean find(int start)方法判断 str 从参数 start 指定位置开始是否有与 regex 匹配的子序列。
⊙ 向指导教师演示程序的运行效果。

6. 实验报告

实验报告的格式如下（可要求学生填写并由实验指导教师签字）：

学号：_____ 班级：_____ 姓名：_____ 时间：_____

实验内容	回答	教师评语
Matcher 类的 find()方法和 lookingAt()方法有何不同？编写程序测试你的结论		

6.3 知识扩展——元词和定位元字符

1. 元词

在进行模式匹配时，Java 允许使用元词，元词代表一种类别，Java 的元词见表 6.1，这些元词代表的意义仅限于 ASCII 表中的字符。

表 6.1 元词

元词	在正则表达式中的写法	意　　义
\p{Lower}	\\p{Lower}	小写字母
\p{Upper}	\\p{Upper}	大写字母
\p{ASCII}	\\p{ASCII}	ASCII 字符
\p{Alpha}	\\p{Alpha}	字母
\p{Digit}	\\p{Digit}	数字字符，即 0～9
\p{Alnum}	\\p{Alnum}	字母或数字
\p{Punct}	\\p{Punct}	标点符号：!"#$%&'()*+,-./:;<=>?@[\]^_`{|}~
\p{Graph}	\\p{Graph}	可视字符：\p{Alnum} \p{Punct}
\p{Print}	\\p{Graph}	可打印字符：\p{Graph}
\p{Blank}	\\p{Blank}	空格或制表符[\t]
\p{Cntrl}	\\p{Cntrl}	控制字符：[\x00-\x1F \x7F]
\p{XDigit}	\\p{XDigit}	十六进制数字：[0-9 a-f A-F]
\p{Space}	\\p{Space}	特殊空格字符：[\t \n \x0B \f \r]

下面的 Example1.java 在进行模式匹配时使用了元词。

Example1.java

```java
import java.util.regex.*;
public class Example1{
    public static void main(String args[]){
        Pattern p;                                              //模式对象
        Matcher m;                                              //匹配对象
        String patternString="\\p{ASCII}\\p{Digit}\\p{Punct}";  //正则表达式
        String matchedString="AA2,BB33%.boy9!";                 //待匹配的字符序列
        p=Pattern.compile(patternString);                       //初始化模式对象
        m=p.matcher(matchedString);                             //用待匹配字符序列初始化匹配对象
        while(m.find()){
            String str=m.group();
            System.out.print("从"+ m.start()+"到"+ m.end()+ "匹配模式子序列: ");
            System.out.println(str);
        }
    }
}
```

2. 定位元字符

定位元字符代表一些字符序列的某些特殊的位置（见表 6.2），如"#.$"代表某行的结尾是字符"#"，"^Dear"代表某行的开头是"Dear"。

表 6.2 定位元字符

元词	在正则表达式中的写法	意　　义
^	^	行首
$	$	行尾
\b	\\b	单词的边界
\B	\\B	非单词的边界
\A	\\A	输入序列的起始部分
\Z	\\Z	输入序列的结尾部分
\G	\\G	前一次匹配模式序列的结尾

在使用定位元字符"^"、"$"时，模式对象应当使用多行标志：

    Matcher p=Pattern.compile(patternString,Pattern.MULTILINE);

下面的 Example2.java 在进行模式匹配时使用了定位元字符。

Example2.java

```java
import java.util.regex.*;
public class Example2 {
    public static void main(String args[]) {
        Pattern p;                                              //模式对象
        Matcher m;                                              //匹配对象
        String patternString="^[0-9].*$";
        String matchedString="1.Java 语言的诞生\n"+ "2.基本数据类型\n"+
                    "3.Java 与 C++ 的关系\n"+ "4.类、对象与接口\n";
```

```
                                          //待匹配的字符序列
p=Pattern.compile(patternString,Pattern.MULTILINE);
m=p.matcher(matchedString);            //用待匹配字符序列初始化匹配对象
while(m.find()) {
    String str=m.group();
    System.out.print("从"+ m.start()+ "到"+ m.end()+ "匹配模式子序列: ");
    System.out.println(str);
patternString="^[0-9].*\\p{Punct}$";
matchedString="1.He paid the shopkeeper some money.\n"+_
             "2.I have already read this book.\n"+_
             "3.How old are you ?\n"+_
             "4.it is worth 2398$";        //待匹配的字符序列
p=Pattern.compile(patternString,Pattern.MULTILINE);
m=p.matcher(matchedString);
while(m.find()) {
    String str=m.group();
    System.out.print("从"+ m.start()+ "到"+ m.end()+ "匹配模式子序列: ");
    System.out.println(str);
}
patternString="\\bun\\p{Alpha}*ed\\b";
matchedString="uned,unship,year,unresolved,unsatisfied unset,unscriped";
p=Pattern.compile(patternString,Pattern.MULTILINE);
m=p.matcher(matchedString);
while(m.find()) {
    String str=m.group();
    System.out.print("从"+ m.start()+ "到"+ m.end()+ "匹配模式子序列: ");
    System.out.println(str);
}
patternString="\\AFrom:.*";
matchedString="From: java@sina.com\n"+ "To:jsp@yahoo.com\n"+ "Subject:hello\n"+_
             "近来好吗?您的上一封信已收到,谢谢您的建议! \n"+_
             "From:java@sina.com\n";
p=Pattern.compile(patternString,Pattern.DOTALL);
m=p.matcher(matchedString);
while(m.find()){
    String str=m.group();
    System.out.print("从"+ m.start()+ "到"+ m.end()+ "匹配模式子序列: ");
    System.out.println(str);
    }
   }
  }
 }
}
```

第7章 常用实用类

本章导读

- 知识概述
- 实验1 比较日期的大小
- 实验2 随机布雷
- 实验3 使用 TreeSet 排序
- 实验4 使用 TreeMap 排序
- 知识扩展——排序和查找、自动装箱和自动拆箱

7.1 知识概述

本章主要讲述了 Date 类、Calendar 类、Math 类、BigInteger 类以及 LinkedList、HashSet、HashMap、TreeSet、TreeMap 等数据结构类。Date 类、Calendar 类处理与时间有关的问题,Math 类提供用来进行数学计算的类方法,而 BigInteger 类可以帮助程序处理特别大的整数。

我们在编写程序时经常要与各种数据打交道,为处理这些数据所选的数据结构对于程序的运行效率是非常重要的。在学习"数据结构"课程的时候,我们要用具体的算法去实现相应的数据结构,如为了实现链表这种数据结构,需要实现向链表中插入节点或从链表中删除节点的算法,感觉有些烦琐。在 JDK 1.2 之后,Java 提供了常见数据结构类:LinkedList、HashSet、HashMap、TreeSet、TreeMap 和 Stack。

7.2 实验练习

7.2.1 比较日期的大小

1. 实验目的

本实验的目的是让学生掌握 Date 类和 Calendar 类的常用方法。

2. 实验要求

编写一个 Java 应用程序,用户从键盘输入两个日期,程序将判断两个日期的大小关系以及两个日期之间的间隔天数。

3. 运行效果示例

运行效果如图 7-1 所示。

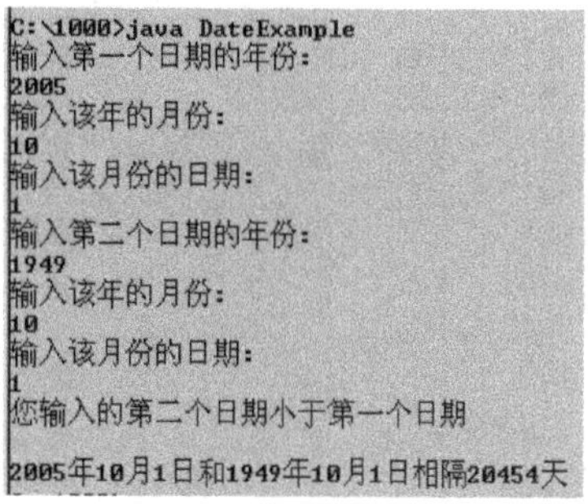

4. 程序模板

按模板要求,将【代码】替换为程序代码。

DateExample.java

```
import java.util.*;
public class DateExample{
    public static void main(String args[]){
        Scanner read=new Scanner(System.in);
        System.out.println("输入第一个日期的年份: ");
        int yearOne=read.nextInt();
        System.out.println("输入该年的月份: ");
        int monthOne=read.nextInt();
        System.out.println("输入该月份的日期: ");
        int dayOne=read.nextInt();
        System.out.println("输入第二个日期的年份: ");
        int yearTwo=read.nextInt();
        System.out.println("输入该年的月份: ");
        int monthTwo=read.nextInt();
        System.out.println("输入该月份的日期: ");
```

图 7-1 Calendar 类的使用

```
        int dayTwo=read.nextInt();
        Calendar calendar=【代码1】        //初始化日历对象
        【代码2】                          //将calendar的时间设置为yearOne年monthOne月dayOne日
        long timeOne=【代码3】             //calendar表示的时间转换成毫秒
        calendar.set(yearTwo,monthTwo-1,dayTwo);
        long timeTwo= calendar.getTimeInMillis();
        Date date1=【代码4】               //用timeOne作为参数构造date1
        Date date2=【代码5】               //用timeTwo作为参数构造date2
        if(date2.equals(date1)){
            System.out.println("两个日期的年、月、日完全相同");
        }
        else if(date2.after(date1)){
            System.out.println("您输入的第二个日期大于第一个日期");
        }
        else if(date2.before(date1)){
            System.out.println("您输入的第二个日期小于第一个日期");
        }
        long 相隔天数=(Math.abs(timeTwo-timeOne))/(1000*60*60*24);
        System.out.printf("%d 年%d 月%d 日和%d 年%d 月%d 日相隔%d 天", yearOne,
                          monthOne,dayOne, yearTwo,monthTwo,dayTwo,相隔天数);
    }
}
```

5. 实验指导与检查

- Calendar 对象设置时间的一个方法是向该方法传递年、月、日。要特别注意【代码2】，整数 0 代表一月，1 代表二月，……，11 代表 12 月。Calendar 对象调用 public long getTimeInMillis()方法可以将时间表示为毫秒，如果运行 Java 程序的本地时区是北京时区，getTimeInMillis()方法返回的是 1970 年 1 月 1 日 8 点至当前时刻的毫秒数。
- 向实验指导教师演示程序的运行效果。

6. 实验报告

实验报告的格式如下（可要求学生填写并由实验指导教师签字）：

学号：_____ 班级：_____ 姓名：_____ 时间：_____

实 验 内 容	回　　答	教 师 评 语
Calendar 对象可以将时间设置到年、月、日、时、分、秒。请改进上面的程序，使用户输入的两个日期包括时、分、秒		
根据本程序中的一些知识，编写一个计算利息（按天计息）的程序，存款的数目和和起止时间从键盘输入		

7.2.2 随机布雷

1. 实验目的

本实验的目的是让学生掌握 LinkedList 类的常用方法。

2. 实验要求

首先编写一个 Block 类，Block 对象具有 String 类型和 boolean 类型的成员变量，Block 对象可以使用 setName(String)方法、getName()方法、isMine()、setBlooean(boolean)方法来设置对象的名字、返回该对象的名字，返回对象的 boolean 类型成员的值、设置对象的 boolean 类型成员的值。

在主类中，要求用一个 Block 类型二维数组模拟 8×8 的方阵，即二维数组的每个单元是一个 Block 对象，然后将二维数组的各单元中的对象存放到一个链表中。

要求在 8×8 的方阵中随机布雷 25 个。

3．运行效果示例

运行效果如图 7-2 所示。

4．程序模板

按模板要求，将【代码】替换为程序代码。

MineExample.java

```
import java.util.*;
class Block {
    String name;
    boolean boo=false;
    public void setName(String name) {
        this.name=name;
    }
    public String getName() {
        return name;
    }
    boolean isMine() {
        return boo
    }
    public void setBoolean(boolean boo) {
        this.boo=boo;
    }
}
public class MineExample{
    public static void main(String args[]) {
        int mine=25;
        Block block[][]=new Block[8][8];
        for(int i=0;i<8;i++){
            for(int j=0;j<8;j++){
                block[i][j]=new Block();
            }
        }
        LinkedList<Block> list=【代码1】         //创建 list
        for(int i=0;i<8;i++) {
            for(int j=0;j<8;j++) {
                【代码2】                        //将 block[i][j]添加到 list 中
            }
        }
        while(mine>=0) {
            int size=【代码3】                   //返回 list 中的节点个数
```

图 7-2 布雷

```java
            int randomIndex=(int)(Math.random()*size);
            Block b=【代码4】            //返回list中索引值为randomIndex的节点中的对象
            b.setName("@");
            b.setBoolean(true);
            list.remove(randomIndex);
            mine--;
        }
        for(int i=0;i<8;i++) {
            for(int j=0;j<8;j++) {
                if(block[i][j].isMine()) {}
                else {
                    int mineNumber=0;
                    for(int k=Math.max(i-1,0);k<=Math.min(i+1,7);k++){
                        for(int t=Math.max(j-1,0);t<=Math.min(j+1,7);t++){
                            if(block[k][t].isMine()){
                                mineNumber++;
                            }
                        }
                    }
                    block[i][j].setName(""+mineNumber);
                }
            }
        }
        for(int i=0;i<8;i++) {
            for(int j=0;j<8;j++) {
                System.out.printf("%2s",block[i][j].getName());
            }
            System.out.printf("%n");
        }
    }
}
```

5．实验指导与检查

向实验指导教师演示程序的运行效果。

6．实验报告

实验报告的格式如下（可要求学生填写并由实验指导教师签字）：

学号：_____ 班级：_____ 姓名：_____ 时间：_____

实 验 内 容	回　答	教师评语
请编写一个应用程序，用一个二维数组模拟4×4方阵，然后将整数1~8随机放入方阵中，要求1~8中的每个数在方阵中恰好出现2次 提示：首先将1~8添加到链表中，使得链表的长度为16，如链表的前8个节点中的数据是1，2，3，4，5，6，7，8，后8个节点中的数据也是1，2，3，4，5，6，7，8；然后随机删除链表中的节点，同时将该节点中的数据顺序地放入方阵中		

7.2.3 使用 TreeSet 排序

1．实验目的

本实验的目的是让学生掌握 TreeSet 类的使用。

2．实验要求

编写一个应用程序，用户从键盘输入 5 个学生的姓名和分数，按成绩排序输出学生的姓名和分数。

3．运行效果示例

运行效果如图 7-3 所示。

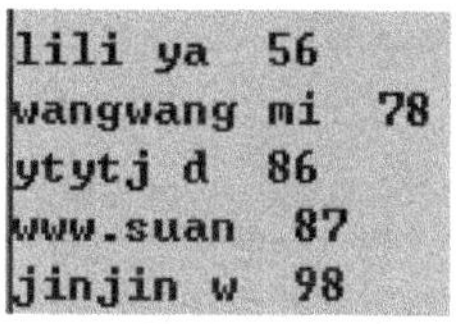

4．程序模板

按模板要求，将【代码】替换为程序代码。

图 7-3 排序（一）

TreeSetExample.java

```java
        import java.util.*;
        public class TreeSetExample {
            public static void main(String args[]) {
                TreeSet<Student> mytree=new TreeSet<Student>();
                for(int i=0;i<5;i++) {
                    Scanner read=new Scanner(System.in);
                    System.out.println("学生的姓名：");
                    String name=read.nextLine();
                    System.out.println("输入分数(整数)：");
                    int score=read.nextInt();
                    【代码1】                            //向 mytree 添加 Student 对象
                }
                Iterator<Student> te=【代码2】          //mytree 返回 Iterator 对象
                while(【代码3】) {                      //判断 te 中是否存在元素
                    Student stu=【代码4】               //返回 te 中的下一个元素
                    System.out.println(""+ stu.name+ "   "+ stu.english);
                }
            }
        }
        class Student implements Comparable {
            int english=0;
            String name;
            Student(int e,String n) {
                english=e;
                name=n;
            }
            public int compareTo(Object b) {
                Student st=(Student)b;
                return (this.english-st.english);
```

5. 实验指导与检查

向实验指导教师演示程序的运行效果。

6. 实验报告

实验报告的格式如下（可要求学生填写并由实验指导教师签字）：

学号：_____ 班级：_____ 姓名：_____ 时间：_____

实 验 内 容	代码完成情况	教 师 评 语
请改写代码，要求按成绩从高到低输出学生的姓名和成绩		

图 7-4 排序（二）

7.2.4 使用 TreeMap 排序

1. 实验目的

本实验的目的是让学生掌握 TreeMap 类的使用。

2. 实验要求

编写一个应用程序，用户从键盘输入 5 个学生的姓名和数学分数、英语分数。程序分别按英语、数学和总分排序输出学生的姓名和分数。

3. 运行效果示例

运行效果如图 7-4 所示。

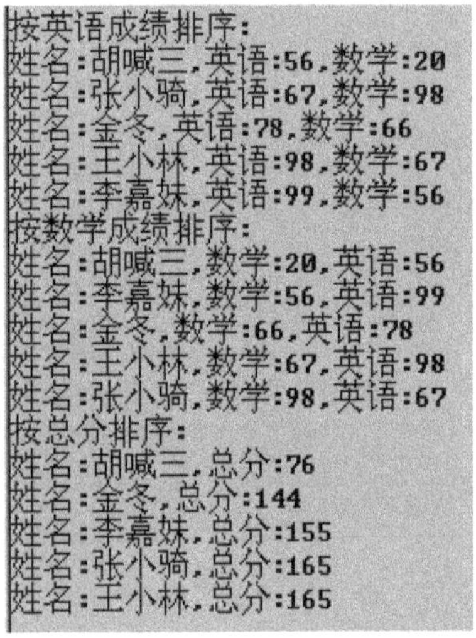

4. 程序模板

按模板要求，将【代码】替换为程序代码。

TreeMapExample.java

```java
import java.util.*;
class MyKey implements Comparable {
    int number=0;
    MyKey(int number) {
        this.number=number;
    }
    public int compareTo(Object b) {
        MyKey st=(MyKey)b;
        if((this.number-st.number)==0) {
            return -1;
        }
        else {
            return (this.number-st.number);
        }
    }
}
```

```
}
class Student {
    String name=null;
    int englishScore,mathScore;
    Student(int e,int m,String name) {
        englishScore=e;
        mathScore=m;
        this.name=name;
    }
}
public class TreeMapExample {
    public static void main(String args[]) {
        TreeMap<MyKey,Student> treemap1=new TreeMap<MyKey,Student>();
        TreeMap<MyKey,Student> treemap2=new TreeMap<MyKey,Student>();
        TreeMap<MyKey,Student> treemap3=new TreeMap<MyKey,Student>();
        for(int i=1;i<=5;i++) {
            Scanner read=new Scanner(System.in);
            System.out.println("学生的姓名: ");
            String name=read.nextLine();
            System.out.println("输入英语分数(整数): ");
            int englishScore=read.nextInt();
            System.out.println("输入数学分数(整数): ");
            int mathScore=read.nextInt();
            Student stu=new Student(englishScore,mathScore,name);
            【代码1】    //向treemap1添加"键-值"对，其中值为stu,要求按英语成绩排序
            【代码2】    //向treemap2添加"键-值"对，其中值为stu,要求按英语成绩排序
            【代码3】    //向treemap3添加"键-值"对，其中值为stu,要求按英语成绩排序
        }
        System.out.println("按英语成绩排序: ");
        Collection<Student> collection=treemap1.values();
        Iterator<Student> iter=collection.iterator();
        while(iter.hasNext()) {
            Student te ==iter.next();
            System.out.printf("姓名:%s,英语:%d,数学:%d\n",te.name,te.englishScore,te.mathScore);
        }
        System.out.println("按数学成绩排序: ");
        collection= treemap2.values();
        iter=collection.iterator();
        while(iter.hasNext()) {
            Student te=iter.next();
            System.out.printf("姓名: %s,数学: %d,英语: %d\n",te.name,te.mathScore,te.englishScore);
        }
        System.out.println("按总分排序: ");
        collection= treemap3.values();
        iter=collection.iterator();
```

```
            while(iter.hasNext()) {
                Student te = iter.next();
                System.out.printf("姓名: %s, 总分: %d\n", te.name, te.englishScore+ te.mathScore);
            }
        }
    }
```

5．实验指导与检查

向实验指导教师演示程序的运行效果。

6．实验报告

实验报告的格式如下（可要求学生填写并由实验指导教师签字）：

学号：_____ 班级：_____ 姓名：_____ 时间：_____

实 验 内 容	代码完成情况	教师评语
请改写代码，使得程序还能按姓名的字典序输出学生的姓名和总分		

7.3 知识扩展——排序和查找、自动装箱和自动拆箱

1．排序和查找

程序可能经常需要对链表按照某种大小关系排序，以便查找一个数据是否与链表中某个节点上的数据相同。java.util 包中的 Collections 类提供了几个类方法（Collections 是类，Collection 是接口），用来处理 Lise<E>接口的数据结构的排序与元素的查找。LinkedList<E>和 ArrayList<E>都是实现 List<E>接口的类，两者的区别是：ArrayList 使用顺序结构存储数据，LiankedList 使用链式结构存储数据。

Collections 类提供的用于排序和查找的类方法如下：

（1）public static sort(List<E> list) ——将 list 中的元素按字典序升序排列。该方法的参数是泛型接口 List<E>，可以传递一个 LinkedList<String>或 ArrayList<String>对象。该方法适合排序字符串数据，即链表中的数据是 String 类型。

（2）public static void sort(List<T> list, Comparator<T> c) ——list 中存储的是实现 Comparable 接口创建的对象，对象通过调用接口方法比较相互之间的大小关系。参数 c 是 Comparator 接口类型，该接口是 java.util 包中的一个接口。compare()方法是接口中的方法，可以向 c 传递一个匿名类的实例，匿名类的方法体必须实现 compare()方法。当 sort()方法对 list 中的元素进行排序时，会调用接口中的 compare()方法对 list 中的数据进行比较。接口回调过程对编程人员是不可见的，Sun 公司在编写 sort()方法时已经实现了这一机制。有时需要查找链表中是否含有与指定数据相等的数据，那么首先要对链表排序，然后使用 binarySearch()方法查找链表中是否含有与指定数据相等的数据。

（3）int binarySearch(List<T> list, T key) ——使用折半法查找 list 是否含有与参数 key 相等的元素，参数 key 将按字典序与 list 中的元素比较是否相等。因此，list 中的元素是 String 类型，list 事先需经过 sort(List<E> list)方法排序。如果找到指定的数据，该方法返回数据在 list 中的顺序位置，否则返回一个负数。

（4）public static int binarySearch(List<T> list, T key, Comparator<T> c) ——按折半法查找 list

是否含有与参数 key 相等的元素，list 必须事先通过 sort(List<T> list, Comparator <T> c)方法排序。

下面的 Example.java 使用了 sort()方法和 binarySearch()方法。

Example.java

```java
import java.util.*;
class Student implements Comparable{
    int score=0;
    String name;
    Student(int s,String n){
        score=s;
        name=n;
    }
    public int compareTo(Object b){    //两个 Student 对象相同当且仅当二者的 score 值相等
        Student st=(Student)b;
        return (this.score-st.score);
    }
}
public class Example{
    public static void main(String args[]){
        LinkedList<String> listOne=new LinkedList<String>();
        listOne.add("bird");
        listOne.add("apple");
        listOne.add("drive");
        listOne.add("fine");
        listOne.add("cap");
        Collections.sort(listOne);
        int number=listOne.size();
        System.out.println("linkTwo 有"+ number+ "个节点: ");
        for(int i=0;i<number;i++){
            String temp=listOne.get(i);
            System.out.println("第"+ i+ "节点: "+ temp);
        }
        String searchedWord="apple";
        int index=Collections.binarySearch(listOne,searchedWord);
        if(index>=0){
            System.out.println("linkOne 中含有与数据"+ searchedWord+ "相同的数据: "+ index);
            String temp=listOne.get(index);
            System.out.println("该数据的信息: "+ temp);
        }
        else{
            System.out.println("linkOne 中不含有与数据"+ searchedWord+ "相同的数据");
        }
        LinkedList<Student> listTwo=new LinkedList<Student>();
        listTwo.add(new Student(92,"Aames"));
        listTwo.add(new Student(62,"Applet"));
```

```java
                listTwo.add(new Student(98,"Abert"));
                listTwo.add(new Student(75,"Balint"));
                listTwo.add(new Student(23,"Zarkit"));
                Collections.sort(listTwo,new Comparator<Student>(){
                                public int compare(Student a,Student b){
                                    return a.compareTo(b);
                                }
                            });
                Student searchedStudent=new Student(75,"javajava");
                index=Collections.binarySearch(listTwo,searchedStudent,new Comparator<Student>(){
                                public int compare(Student a,Student b){
                                    return a.compareTo(b);
                                }
                            });
                if(index>=0){
                    System.out.println("linkTwo 中含有与数据"+ searchedStudent.name+ ","+
                                    searchedStudent.score+ "相同的数据: "+ index);
                    Student temp=listTwo.get(index);
                    System.out.println("该数据的信息: "+ temp.name+ ","+ temp.score);
                }
                else{
                    System.out.println("linkTwo 中不含有与数据"+ searchedStudent.name+
                                    ","+ searchedStudent.score+ "相同的数据");
                }
                number=listTwo.size();
                System.out.println("linkTwo 有"+ number+ "个节点:");
                for(int i=0;i<number;i++){
                    Student temp=listTwo.get(i);
                    System.out.printf("第"+ i+ "节点: %s,%d\n",temp.name,temp.score);
                }
            }
        }
```

2. 洗牌和旋转

Collections 类还提供了将链表中的数据重新随机排列的类方法和旋转链表中数据的类方法。

（1） public static void shuffle(List<E> list) ——将 list 中的数据重新随机排列。

（2） static void rotate(List<E> list, int distance) ——旋转链表中的数据。使用该方法后，list 索引 i 处的数据将是该方法调用前 list 索引 (i−distance) mod list.size() 处的数据。例如，假设 list 的数据依次为 t、a、n、k、s，那么，在 Collections.rotate(list, 1) 之后，list 的数据依次为 s、t、a、n、k。当方法的参数 distance 取正值时，向右转动 list 中的数据，取负值时，向左转动 list 中的数据。

（3） public static void reverse(List<E> list) ——翻转 list 中的数据。假设 list 索引处的数据依次为 t、a、n、k、s，那么，在 Collections. reverse (list) 方法之后，list 索引处的数据依次为 s、k、n、a、t。

下面的 Ex.java 使用了 shuffle() 方法、reverse() 方法和 rotate() 方法。

Ex.java
```java
import java.util.*;
public class Ex{
    public static void main(String args[]){
        LinkedList<Integer> list1=new LinkedList<Integer>();
        for(int i=1;i<=16;i++){
            list1.add(new Integer(i));
        }
        Collections.shuffle(list1);
        Iterator<Integer> iter1=list1.iterator();
        while(iter1.hasNext()){
            Integer te=iter1.next();
            System.out.printf("%3d",te.intValue() );
        }
        LinkedList<String> list2=new LinkedList<String>();
        list2.add("e");
        list2.add("d");
        list2.add("c");
        list2.add("b");
        list2.add("a");
        Collections.reverse(list2);
        Iterator<String> iter2=list2.iterator();
        System.out.printf("\n");
        while(iter2.hasNext()){
            String te=iter2.next();
            System.out.printf("%2s",te);
        }
        for(int k=1;k<=5;k++){
            Collections.rotate(list2,-1);
            iter2=list2.iterator();
            System.out.printf("\n");
            while(iter2.hasNext()){
                String te=iter2.next();
                System.out.printf("%2s",te);
            }
        }
    }
}
```

3. 自动装箱和自动拆箱

对于数据结构类,SDK 1.5 为其增加了基本类型数据和引用型数据相互自动转换的功能,称为基本数据类型的自动装箱和自动拆箱(Autoboxing and Auto-Unboxing of Primitive Type)。

在没有自动装箱和自动拆箱功能之前,我们不能将 int 类型数据添加到类似链表的数据结构中。SDK 1.5 后,程序允许把一个 int 类型数据添加到链表等数据结构中,系统会自动完成基本类型到引用类型的转换,这一过程称为自动装箱。当从一个数据结构中获取含有的基本数据类型时,系统自动完成引用类型到基本类型的转换,这一过程称为自动拆箱。下面的 Example.java 中使用了自动装箱和自动拆箱。

Example.java

```java
import java.util.*;
public class Example {
    public static void main(String args[]) {
        ArrayList<Integer> list=new ArrayList<Integer>();
        for(int i=0;i<10;i++) {
            list.add(i);                         //自动装箱
        }
        for(int k=list.size()-1;k>=0;k--) {
            int m=list.get(k);                   //自动拆箱
            System.out.printf("%3d",m);
        }
    }
}
```

第 8 章

多 线 程

本章导读

- ✿ 知识概述
- ✿ 实验 1 Thread 类的子类创建线程
- ✿ 实验 2 使用 Thread 类创建线程
- ✿ 实验 3 吵醒休眠的线程
- ✿ 实验 4 排队买票
- ✿ 实验 5 线程联合
- ✿ 知识扩展——Timer 类和 TimerTask 类

8.1 知识概述

可以使用 Thread 类或 Thread 类的子类创建线程对象。线程是比进程更小的执行单位。一个进程在其执行过程中可以产生多个线程，形成多条执行线索。每条线索，即每个线程，也有它自身的产生、存在和消亡的过程，也是一个动态的概念。线程在它的一个完整的生命周期中通常要经历 4 种状态：新建、运行、中断和死亡。

Java 虚拟机（JVM）中的线程调度器负责管理线程，在采用时间片的系统中，每个线程都有机会获得 CPU 的使用权。当线程使用 CPU 资源的时间终了时，即使线程没有完成自己的全部操作，Java 调度器也会中断当前线程的执行，把 CPU 的使用权切换给下一个排队等待的线程，当前线程将等待 CPU 资源的下一次轮回，然后从中断处继续执行。Java 虚拟机中的线程调度器把线程的优先级分为 10 个级别，分别用 Thread 类中的类常量表示，在实际编程时，不提倡使用线程的优先级来保证算法的正确执行。要编写正确的、跨平台的多线程代码，必须假设线程在任何时间都有可能被剥夺 CPU 资源的使用权。

线程同步是很重要的内容之一，在处理线程同步时，要做的第一件事就是要把修改数据的方法用关键字 synchronized 来修饰。一个方法使用关键字 synchronized 修饰后，如果一个线程 A 占有 CPU 资源期间，使得该方法被调用执行，那么在该同步方法返回之前，即同步方法调用执行完毕之前，其他占有 CPU 资源的线程一旦调用这个同步方法就会引起堵塞，堵塞的线程要一直等到堵塞的原因消除（同步方法返回），再排队等待 CPU 资源，以便使用这个同步方法。

在处理同步问题时，有时会涉及一些特殊问题，如当一个线程使用的同步方法中用到某个变量，而此变量又需要其他线程修改后才能符合本线程的需要。为了解决这样的问题，正在使用同步方法的线程可以执行 wait()方法，可以中断线程的执行，使本线程等待，暂时让出 CPU 的使用权，并允许其他线程使用这个同步方法。其他线程如果在使用这个同步方法时不需要等待，那么它使用完这个同步方法的同时，应当用 notifyAll()方法通知所有的由于使用这个同步方法而处于等待的线程结束等待。曾中断的线程就会重新排队等待 CPU 资源，以便从刚才的中断处继续执行这个同步方法。

线程联合也是编写多线程问题中经常使用的技术，一个线程 A 在占有 CPU 资源期间，可以让线程 B 和线程 A 联合 B.join()，称 A 在运行期间联合了 B。如果线程 A 在占有 CPU 资源期间一旦联合线程 B，那么线程 A 将立刻中断执行，一直等到它联合的线程 B 执行完毕，线程 A 再重新排队等待 CPU 资源，以便恢复执行。如果线程 A 准备联合的线程 B 已经结束或没有就绪排队等待 CPU 资源，那么 B.join()不会产生任何效果。

线程分为守护（Daemon）线程和非守护线程，非守护线程也称为用户线程。一个线程调用 setDaemon(boolean on)方法可以将自己设置成一个守护线程，如 thread.setDaemon(true)。

当程序中的所有用户线程都已结束运行时，即使守护线程的 run()方法中还有需要执行的语句，守护线程也立刻结束运行。

8.2 实验练习

8.2.1 使用 Thread 的子类创建线程

1．实验目的

本实验的目的是让学生掌握用 Thread 类的子类创建线程的方法。

2．实验要求

编写一个 Java 应用程序，在主线程中再创建两个线程，一个线程负责给出键盘上字母键上的字母，另一个线程负责让用户在命令行输入所给出的字母。

3．运行效果示例

运行效果如图 8-1 所示。

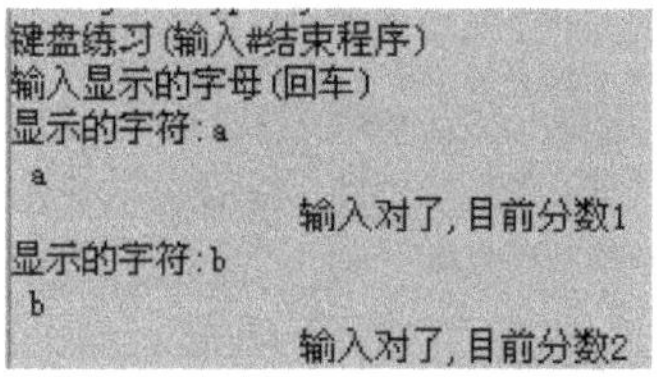

4．程序模板

按模板要求，将【代码】替换为程序代码。

图 8-1 按键练习

TypeKey.java

```
public class TypeKey{
    public static void main(String args[]){
        System.out.println("键盘练习(输入#结束程序)");
        System.out.printf("输入显示的字母(回车)\n");
        Letter letter;
        letter=new Letter();
        GiveLetterThread giveChar;
        InuptLetterThread typeChar;
        【代码1】                              //创建线程 giveChar
        giveChar.setLetter(letter);
        giveChar.setSleepLength(3200);
        【代码2】                              //创建线程 typeChar
        typeChar.setLetter(letter);
        giveChar.start();
        typeChar.start();
    }
}
```

Letter.java

```
public class Letter{
    char c ='\0';
    public void setChar(char c){
        this.c=c;
    }
    public char getChar(){
        return c;
    }
}
```

GiveLetterThread.java

```java
public class GiveLetterThread extends Thread{
    Letter letter;
    char startChar='a',endChar='z';
    int sleepLength=5000;
    public void setLetter(Letter letter){
        this.letter=letter;
    }
    public void setSleepLength(int n){
        sleepLength=n;
    }
    public void run(){
        char c=startChar;
        while(true){
            letter.setChar(c);
            System.out.printf("显示的字符: %c\n",letter.getChar());
            try{
                【代码3】            //调用 sleep()方法,使得线程中断 sleepLength 毫秒
            }
            catch(InterruptedException e){ }
            c=(char)(c+1);
            if(c>endChar){
                c=startChar;
            }
        }
    }
}
```

InuptLetterThread.java

```java
import java.awt.*;
import java.util.Scanner;
public class InuptLetterThread extends Thread{
    Scanner reader;
    Letter letter;
    int score=0;
    InuptLetterThread(){
        reader=new Scanner(System.in);
    }
    public void setLetter(Letter letter){
        this.letter=letter;
    }
    public void run(){
        while(true){
            String str=reader.nextLine();
            char c=str.charAt(0);
            if(c==letter.getChar()){
```

```
                score++;
                System.out.printf("\t\t 输入对了，目前分数%d\n",score);
            }
            else{
                System.out.printf("\t\t 输入错了，目前分数%d\n",score);
            }
            if(c=='#'){
                System.exit(0);
            }
        }
    }
}
```

5．实验指导与检查

- 使用 Thread 类的子类创建线程一定要重写父类的 run()方法，否则线程什么也不做。
- 线程的 run()方法开始执行后，不要让线程再调用 start()方法。
- 向实验指导教师演示程序的运行效果。

6．实验报告

实验报告的格式如下（可要求学生填写并由实验指导教师签字）：

学号：_____ 班级：_____ 姓名：_____ 时间：_____

实 验 内 容	回　　答	教 师 评 语
改进 GiveLetterThread 类，使得该类创建的线程能让用户熟练使用更多的键		

8.2.2 使用 Thread 类创建线程

1．实验目的

本实验的目的是让学生学习用 Thread 类创建线程，掌握哪些数据是线程之间共享的，哪些线程是线程独有的。

2．实验要求

编写一个 Java 应用程序，在主线程中用 Thread 类再创建 2 个线程，这 2 个线程共享一个 int 类型的数据，并有自己独占的数据。

3．运行效果示例

运行效果如图 8-2 所示。

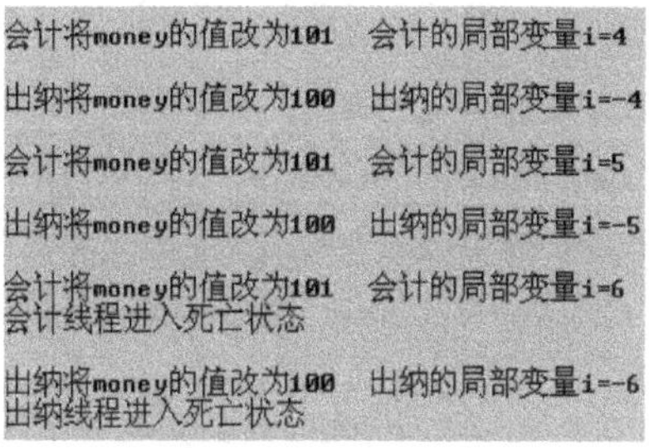

4．程序模板

按模板要求，将【代码】替换为程序代码。

BankExample.java

```
class Bank implements Runnable{
    int money=100;
    Thread zhang,keven;
```

图 8-2　线程间共享数据

```
            Bank(){
                zhang=new Thread(this);
                zhang.setName("会计");
                keven=new Thread(this);
                keven.setName("出纳");
            }
            public void run(){
                int i =0 ;
                while(true) {
                    if(【代码1】){                    //判断线程 zhang 是否正在占有 CPU 资源
                        i=i+ 1;
                        money=money+ 1;
                        System.out.printf("\n%s 将 money 的值改为%d\t",zhang.getName(),money);
                        System.out.printf("%s 的局部变量 i=%d\n",zhang.getName(),i);
                        if(i>=6){
                            System.out.printf("%s 线程进入死亡状态\n",zhang.getName());
                            【代码2】                    //使得线程 zhang 进入死亡状态
                        }
                        try{
                            Thread.sleep(1000);
                        }
                        catch(InterruptedException e){ }
                    }
                    else if(【代码3】){                //判断线程 keven 是否正在占有 CPU 资源
                        i=i-1;
                        money=money-1;
                        System.out.printf("\n%s 将 money 的值改为%d\t",keven.getName(),money);
                        System.out.printf("%s 的局部变量 i=%d\n",keven.getName(),i);
                        if(i<=-6){
                            System.out.printf("%s 线程进入死亡状态\n",keven.getName());
                            【代码4】                    //使得线程 keven 进入死亡状态
                        }
                        try{
                            Thread.sleep(1000);
                        }
                        catch(InterruptedException e){ }
                    }
                }
            }
        }
        public class BankExample{
            public static void main(String args[]){
                Bank bank=new Bank();
                bank.zhang.start();
                bank.keven.start();
            }
        }
```

5. 实验指导与检查

- 对于构造方法 Thread(Runnable target)创建的线程，轮到它来享用 CPU 资源时，目标对象就会自动调用接口中的 run()方法。因此，对于使用同一目标对象的线程，目标对象的成员变量自然就是这些线程共享的数据单元。
- 对于具有相同目标对象的线程，当其中一个线程享用 CPU 资源时，目标对象自动调用接口中的 run()方法。这时，run()方法中的局部变量被分配内存空间。当轮到另一个线程享用 CPU 资源时，目标对象会再次调用接口中的 run()方法，那么 run()方法中的局部变量会再次分配内存空间。不同线程的 run()方法中的局部变量互不干扰，一个线程改变了自己的 run()方法中局部变量的值不会影响其他线程的 run()方法中的局部变量。
- 向实验指导教师演示程序的运行效果。

6. 实验报告

实验报告的格式如下（可要求学生填写并由实验指导教师签字）：

学号：_____ 班级：_____ 姓名：_____ 时间：_____

实 验 内 容	回　　答	教师评语
在实验代码的基础上，再增加一个线程，该线程的名字是 xiaoming		

8.2.3 吵醒休眠的线程

1. 实验目的

本实验的目的是让学生掌握线程的 interrupt()方法。

2. 实验要求

编写一个 Java 应用程序，在主线程中有 3 个线程：zhangWorker、wangWorker 和 boss。线程 zhangWorker 和 wangWorker 分别负责在命令行输出"搬运苹果"和"搬运香蕉"，这两个线程分别各自输出 20 行，每输出一行信息就准备休息 10 秒钟，但是 boss 线程负责不让 zhangWorker 和 wangWorker 休息。

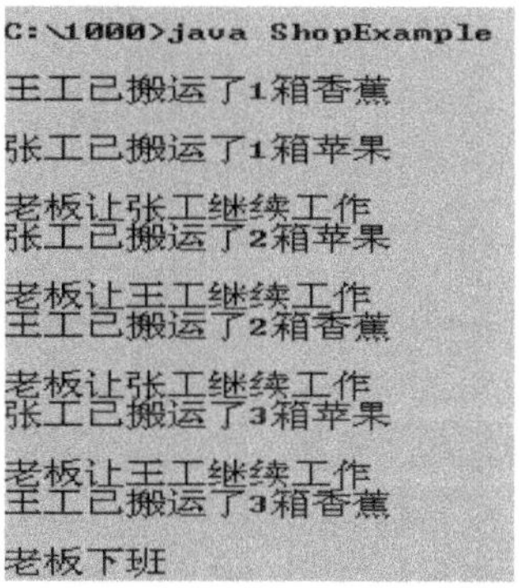

3. 运行效果示例

运行效果如图 8-3 所示。

4. 程序模板

按模板要求，将【代码】替换为 Java 程序代码。

ShopExample.java

```
class Shop implements Runnable{
    Thread zhangWorker,wangWorker,boss;
    Shop(){
        boss=new Thread(this);
        zhangWorker=new Thread(this);
        wangWorker=new Thread(this);
        zhangWorker.setName("张工");
        wangWorker.setName("王工");
```

图 8-3 吵醒线程

```
                boss.setName("老板");
            }
            public void run(){
                int i=0;
                if(Thread.currentThread()==zhangWorker){
                    while(true){
                        try{
                            i++;
                            System.out.printf("\n%s 已搬运了%d 箱苹果\n",zhangWorker.getName(),i);
                            if(i==3){
                                return;
                            }
                            【代码 1】                              //zhangWorker 休眠 10 秒（10000 毫秒）
                        }
                        catch(InterruptedException e){
                            System.out.printf("\n%s 让%s 继续工作",boss.getName(),zhangWorker.getName());
                        }
                    }
                }
                else if(Thread.currentThread()==wangWorker){
                    while(true){
                        try{
                            i++;
                            System.out.printf("\n%s 已搬运了%d 箱香蕉\n",wangWorker.getName(),i);
                            if(i==3){
                                return;
                            }
                            【代码 2】                              //wangWorker 休眠 10 秒（10000 毫秒）
                        }
                        catch(InterruptedException e){
                            System.out.printf("\n%s 让%s 继续工作",boss.getName(),wangWorker.getName());
                        }
                    }
                }
                else if(Thread.currentThread()==boss){
                    while(true){
                        【代码 3】                                  //吵醒 zhangWorker
                        【代码 4】                                  //吵醒 wangWorker
                        if(!(wangWorker.isAlive() || zhangWorker.isAlive())){
                            System.out.printf("%n%s 下班",boss.getName());
                            return;
                        }
                    }
                }
            }
        }
        public class ShopExample{
```

```
        public static void main(String args[]){
            Shop shop=new Shop();
            shop.zhangWorker.start();
            shop.wangWorker.start();
            shop.boss.start();
        }
    }
```

5. 实验指导与检查

- intertupt()方法经常用来"吵醒"休眠的线程。当一些线程调用 sleep()方法处于休眠状态时，一个使用 CPU 资源的其他线程在执行过程中，可以让休眠的线程分别调用 interrupt()方法"吵醒"自己，即导致休眠的线程发生 InterruptedException 异常，从而结束休眠，重新排队等待 CPU 资源。
- 向实验指导教师演示程序的运行效果。

6. 实验报告

实验报告的格式如下（可要求学生填写并由实验指导教师签字）：

学号：_____ 班级：_____ 姓名：_____ 时间：_____

实 验 内 容	回　　答	教 师 评 语
反复运行上述程序，观察每次程序运行后的输出结果，如果出现两次运行的结果不尽相同。请解释其中的原因		

8.2.4 排队买票

1. 实验目的

本实验的目的是让学生掌握怎样处理多线程中的同步问题，学会使用 wait()、notify()和 notifyAll()方法。

2. 实验要求

编写一个 Java 应用程序，模拟 5 个人排队买票。售票员只有 1 张五元的钱，电影票五元钱一张。假设 5 个人的名字及排队顺序是：赵、钱、孙、李、周。"赵"拿 1 张二十元的人民币买 2 张票，"钱"拿 1 张二十元的人民币买 1 张票，"孙"拿 1 张十元的人民币买 1 张票，"李"拿 1 张十元的人民币买 2 张票，"周"拿 1 张五元的人民币买 1 张票。要求售票员按如下规则找赎：

- 二十元买 2 张票，找零：1 张十元；不许找零：2 张五元。
- 二十元买 1 张票，找零：1 张十元，1 张五元；不许找零：3 张五元。
- 十元买 1 张票，找零：1 张五元。

3. 运行效果示例

运行效果如图 8-4 所示。

4. 程序模板

按模板要求，将【代码】替换为程序代码。

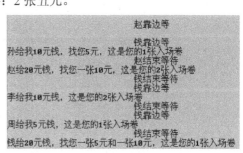

图 8-4　线程排队

SaleExample.java

```java
class TicketSeller{                                  //负责卖票的类
    int fiveNumber=1,tenNumber=0,twentyNumber=0;
    public synchronized void sellTicket(int receiveMoney,int buyNumber){
        if(receiveMoney==5){
            fiveNumber=fiveNumber+1;
            System.out.printf("\n%s 给我 5 元钱，这是您的 1 张入场券",Thread.currentThread().getName());
        }
        else if(receiveMoney==10&&buyNumber==2){
            tenNumber=tenNumber+1;
            System.out.printf("\n%s 给我 10 元钱，这是您的 2 张入场券",
                               Thread.currentThread().getName());
        }
        else if(receiveMoney==10&&buyNumber==1){
            while(【代码 1】){                        //给出线程需等待的条件
                try{
                    System.out.printf("\n%30s 靠边等",Thread.currentThread().getName());
                    【代码 2】                        //线程进入等待状态
                    System.out.printf("\n%30s 结束等待\n",Thread.currentThread().getName());
                }
                catch(InterruptedException e){ }
            }
            fiveNumber=fiveNumber-1;
            tenNumber=tenNumber+1;
            System.out.printf("\n%s 给我 10 元钱，找您 5 元，这是您的 1 张入场券",
                               Thread.currentThread().getName());
        }
        else if(receiveMoney==20&&buyNumber==1){
            while(【代码 3】){                        //给出线程需等待的条件
                try{
                    System.out.printf("\n%30s 靠边等",Thread.currentThread().getName());
                    【代码 4】                        //线程进入等待状态
                    System.out.printf("\n%30s 结束等待",Thread.currentThread().getName());
                }
                catch(InterruptedException e){ }
            }
            fiveNumber=fiveNumber-1;
            tenNumber=tenNumber-1;
            twentyNumber=twentyNumber+1;
            System.out.printf("\n%s 给 20 元钱，找您一张 5 元和一张 10 元，这是您的 1 张入场券",
                               Thread.currentThread().getName());
        }
        else if(receiveMoney==20&&buyNumber==2){
            while(【代码 5】){                        //给出线程需等待的条件
                try{
                    System.out.printf("\n%30s 靠边等\n",Thread.currentThread().getName());
                    【代码 6】                        //线程进入等待状态
                    System.out.printf("\n%30s 结束等待",Thread.currentThread().getName());
                }
                catch(InterruptedException e){ }
```

```
        }
            tenNumber=tenNumber-1;
            twentyNumber=twentyNumber+1;
            System.out.printf("\n%s 给 20 元钱，找您一张 10 元，这是您的 2 张入场券",
                                        Thread.currentThread().getName());
        }
        【代码 7】                                    //通知等待的线程结束等待
    }
}
class Cinema implements Runnable{                   //实现 Runnable 接口的类
    Thread zhao,qian,sun,li,zhou;                   //电影院中买票的线程
    TicketSeller seller;                            //电影院的售票员
    Cinema(){
        zhao=new Thread(this);
        qian=new Thread(this);
        sun=new Thread(this);
        li=new Thread(this);
        zhou=new Thread(this);
        zhao.setName("赵");
        qian.setName("钱");
        sun.setName("孙");
        li.setName("李");
        zhou.setName("周");
        seller=new TicketSeller();
    }
    public void run()
        if(Thread.currentThread()==zhao){
            seller.sellTicket(20,2);
        }
        else if(Thread.currentThread()==qian){
            seller.sellTicket(20,1);
        }
        else if(Thread.currentThread()==sun){
            seller.sellTicket(10,1);
        }
        else if(Thread.currentThread()==li){
            seller.sellTicket(10,2);
        }
        else if(Thread.currentThread()==zhou){
            seller.sellTicket(5,1);
        }
    }
}
public class SaleExample{
    public static void main(String args[]){
        Cinema cinema=new Cinema();
        cinema.zhao.start();
        try{
            Thread.sleep(1000);
        }
        catch(InterruptedException e){ }
```

```
            cinema.qian.start();
            try{
                Thread.sleep(1000);
            }
            catch(InterruptedException e){ }
            cinema.sun.start();
            try{
                Thread.sleep(1000);
            }
            catch(InterruptedException e){ }
            cinema.li.start();
            try{
                Thread.sleep(1000);
            }
            catch(InterruptedException e){ }
            cinema.zhou.start();
        }
    }
```

5. 实验指导与检查

- 当一个线程使用的同步方法中用到某个变量，而此变量又需要其他线程修改后才能符合本线程的需要时，可以在同步方法中使用 wait()方法。使用 wait()方法可以中断方法的执行，使本线程等待，暂时让出 CPU 的使用权，并允许其他线程使用这个同步方法。
- 可以用 notifyAll()方法通知所有的由于使用同步方法而处于等待的线程结束等待，曾中断的线程就会重新排队等待 CPU 资源，以便从刚才的中断处继续执行这个同步方法，也可以使用 notify()方法通知处于等待线程中的某一个结束等待。
- 向实验指导教师演示程序的运行效果。

6. 实验报告

实验报告的格式如下（可要求学生填写并由实验指导教师签字）：

学号:_____   班级:_____   姓名:_____   时间:_____

实 验 内 容	回　　答	教 师 评 语

将程序中的
　　while(【代码】){　　　//给出线程需等待的条件
　　　……
　　}
改写成：
　　if(【代码】){　　　//给出线程需等待的条件
　　　……
　　}
是否合理？说明理由		

8.2.5　线程联合

1. 实验目的

本实验的目的是让学生掌握线程联合的方法。

2. 实验要求

编写一个 Java 应用程序，在主线程中再创建 3 个线程："运货司机"、"装运工"和"仓库管理员"。要求线程"运货司机"在占有 CPU 资源后立刻联合线程"装运工"，也就是让"运货司机"一直等到"装运工"完成工作才能开车，而"装运工"在占有 CPU 资源后立刻联合线程"仓库管理员"，也就是让"装运工"一直等到"仓库管理员"打开仓库才能开始搬运货物。

3. 运行效果示例

运行效果如图 8-5 所示。

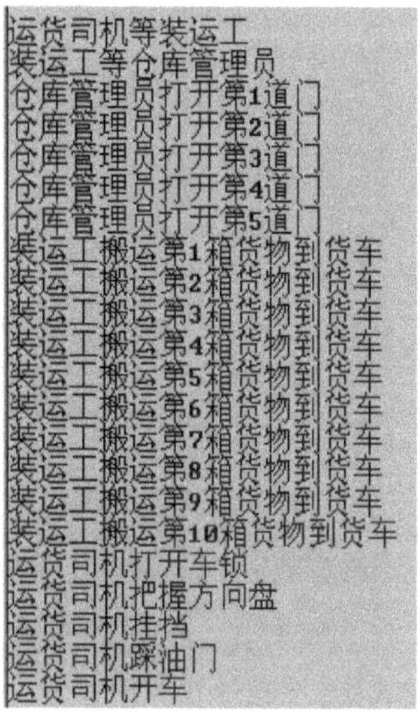

4. 程序模板

按模板要求，将【代码】替换为程序代码。

JoinExample.java

```
class JoinThread implements Runnable{
    Thread 运货司机,装运工,仓库管理员;
    String step[]={"打开车锁","把握方向盘","挂挡","踩油门","开车"};
    JoinThread(){
        运货司机=new Thread(this);
        装运工=new Thread(this);
        仓库管理员=new Thread(this);
        运货司机.setName("运货司机");
        装运工.setName("装运工");
        仓库管理员.setName("仓库管理员");
    }
    public void run(){
        if(Thread.currentThread()==运货司机){
            System.out.printf("\n%s 等%s",运货司机.getName(),装运工.getName());
            try{
                【代码1】                        //占有 CPU 资源期间联合线程：装运工
            }
            catch(InterruptedException e){ }
            for(int i=0;i<step.length;i++){
                System.out.printf("\n%s%s",运货司机.getName(),step[i]);
                try{
                    运货司机.sleep(500);
                }
                catch(InterruptedException ee){ }
            }
        }
        else if(Thread.currentThread()==装运工){
            System.out.printf("\n%s 等%s",装运工.getName(),仓库管理员.getName());
            try{
                【代码2】                        //占有 CPU 资源期间联合线程：仓库管理员
            }
            catch(InterruptedException e){ }
            for(int i=1;i<=10;i++){
                System.out.printf("\n%s 搬运第%d 箱货物到货车",装运工.getName(), i);
                try{
                    装运工.sleep(1500);
                }
```

图 8-5 线程联合

```
                catch(InterruptedException ee){ }
            }
        }
        else if(Thread.currentThread()==仓库管理员){
            for(int i=1;i<=5;i++){
                System.out.printf("\n%s 打开第%d 道门",仓库管理员.getName(),i);
                try{
                    仓库管理员.sleep(1000);
                }
                catch(InterruptedException e){ }
            }
        }
    }
}
public class JoinExample{
    public static void main(String args[]){
        JoinThread a=new JoinThread();
        a.运货司机.start();
        a.装运工.start();
        a.仓库管理员.start();
    }
}
```

5．实验指导与检查

- 一个线程 A 在占有 CPU 资源期间，可以让其他线程调用 join()方法与本线程联合，则称 A 在运行期间联合了 B。线程 A 在占有 CPU 资源期间一旦联合 B 线程，那么线程 A 将立刻中断执行，一直等到它联合的线程 B 执行完毕，线程 A 再重新排队等待 CPU 资源，以便恢复执行。如果 A 准备联合的线程 B 还没有就绪排队或已经结束，那么 B.join()不会产生任何效果。
- 向实验指导教师演示程序的运行效果。

6．实验报告

实验报告的格式如下（可要求学生填写并由实验指导教师签字）：

学号：_____ 班级：_____ 姓名：_____ 时间：_____

实 验 内 容	回 答	教 师 评 语
在上述程序中再增加一个线程："公司老板"。要求"仓库管理员"在打开门之前，必须经过"公司老板"同意。"公司老板"讲完 3 句话："考虑一下"、"我查一下"、"好，可以开门"之后，"仓库管理员"开始开门		

8.3　知识扩展——Timer 类和 TimerTask 类

Timer 与 TimerTask 类可以用来安排将来某个时间要执行的任务，尽管可以使用线程知识来安排将来某个时间要执行的任务，但是使用 Timer 类和 TimerTask 类可以简化这一过程。

1．任务的创建

TimerTask 类负责创建将来某个时间要执行的任务。这里称 TimerTask 类创建的对象为一个任务，可以使用构造方法 TimerTask()创建一个对象。TimerTask 类实现了 Runnable 接口，因此应当编写一个 TimerTask 类的子类，重写 run()方法，用子类创建任务对象。当该任务对象被通知执行时，任务对象将自动调用 run()方法，完成自己的工作。

2．任务的安排和执行

Timer 类负责任务的安排和执行的时间。Timer 对象可以安排一个 TimerTask 对象什么时间开始运行，当指定的时间到来时，将自动启动一个用户线程或守护线程。在该线程中，TimerTask 对象调用 run()方法完成工作。可以使用 Time 类的无参数构造方法 Timer()创建一个 Timer 对象，该对象将启动一个用户线程来完成它安排的任务。也可以使用 Time 类的有参数构造方法 Timer(boolean isDaemon)创建一个 Timer 对象，如果 isDaemon 取值为 true，该对象将启动一个守护线程来完成它安排的任务。

Timer 对象可以调用如下方法安排将来某个时间要执行的任务。

（1）void schedule(TimerTask task, Date time) ——安排的任务 task 在时间 time 执行，如果 time 已经是过去的时间，该任务 task 立刻被执行。

（2）public void schedule(TimerTask task, Date firstTime,long period) ——安排的任务 task 在时间 firstTime 首次执行。当到达时间 firstTime 时，该任务首次执行，然后该任务被不断地间隔重复执行，间隔的时间由参数 period 和任务执行花费的时间长短来决定（单位为毫秒）。也就是说，在上一次执行任务执行完后，再延时 period 毫秒重复执行该任务，由于每次执行任务花费的时间不尽相同，因此间隔的时间也不尽相同。如果 firstTime 已经是过去的时间，那么让该任务立刻执行，并根据当前时间与 firstTime 的差计算出隔按次数，根据该次数将该任务反复执行若干次后，再按时间间隔重复执行该任务。

（3）public void schedule(TimerTask task, long delay) ——该方法一旦被调用，delay 毫秒后 task 执行。

（4）public void schedule(TimerTask task, long delay,long period) ——该方法一旦被调用，delay 毫秒后 task 首次执行，然后该任务被不断地间隔重复执行，间隔的时间由参数 period 和任务执行花费的时间长短来决定（单位为毫秒）。也就是说，在上一次执行任务执行完后，再延时 period 毫秒重复执行该任务。

（5）public void scheduleAtFixedRate(TimerTask task, Date firstTime, long period) ——安排的任务 task 在时间 firstTime 首次执行，然后该任务被不断地间隔重复执行，间隔的时间由参数 period 决定（单位为毫秒），间隔时间不依赖于上次任务执行所花费的时间，只与上一次任务开始执行的时间有关。也就是说，在上一次任务开始执行后，再延时 period 毫秒重复执行该任务，这样每次间隔的时间是完全相同的。如果因为某种原因，某次任务的执行花费了过多的时间，就可能发生该次任务没有执行完毕下一次任务又开始执行的追逐现象。如果 firstTime 已经是过去的时间，那么该任务 task 立刻执行，并根据当前时间与 firstTime 的差计算出隔按次数。根据该次数，将该任务反复执行若干次后，再按时间间隔重复执行该任务。

（6）public void scheduleAtFixedRate(TimerTask task, long delay, long period) ——该方法一旦被调用，delay 毫秒后 task 首次执行，然后该任务被不断地间隔重复执行，间隔的时间由参数 period 决定（单位为毫秒），不依赖于上一次任务执行所花费的时间，只与上一次任务开始执行的时间有关。也就是说，在上一次任务开始执行后，再延时 period 毫秒重复执行该任务，这样每次间隔的时间是完全相同的。如果因为某种原因，某次任务的执行花费了过多的时间，就可能发生该次任务没有执行完毕下一次任务又开始执行的追逐现象。

3. 任务的取消

有两种办法取消任务的执行。

第一种办法是让 TimerTask 对象调用 cancel()方法。如果该任务还没有被 Timer 对象安排执行，那么 Timer 对象将不能安排该任务；如果该任务已经被 Timer 对象安排执行，但还没有开始执行，该任务将不会被执行；如果该任务已经被 Timer 对象安排执行，并且已经开始执行，那么本次执行完之后，该任务将不再重复执行。

第二种办法是让 Timer 调用 cancel()方法取消 Timer 对象安排的所有任务。如果安排的任务还没有开始执行，这些任务将不会被执行；如果安排的某些任务已经开始执行，这些任务将不会重复执行。

下面的 Example.java 使用了 Timer 类和 TimerTask 类。

Example.java

```java
import java.util.*;
class WorkTask extends TimerTask{
    int i=0;
    public void run(){
        i++;
        System.out.print("*");
        if(i>=12){
            System.out.print("END");
            cancel();
        }
    }
}
class WordTask extends TimerTask{
    public void run(){
        System.out.printf("A");
    }
}
public class Example{
    public static void main(String args[]){
        WorkTask workTask=new WorkTask();
        WordTask wordTask=new WordTask();
        Timer boss=new Timer();
        boss.schedule(wordTask,1000,1000);
        Calendar calendar= Calendar.getInstance();
        calendar.set(2012,9,1,8,0,30);          //时间设置为2012年9月1日8时0分30秒
        long start=calendar.getTimeInMillis();
        Date date=new Date(start);
        boss.scheduleAtFixedRate(workTask,date,2000);
        System.out.println("任务布置完毕");
        calendar.set(2012,9,1,8,20,50);
        long end=calendar.getTimeInMillis();
        try{
            Thread.sleep(end-start);
        }
        catch(InterruptedException e){ }
        boss.cancel();
    }
}
```

第 9 章

输入流和输出流

本章导读

- ✿ 知识概述
- ✿ 实验 1 文件加密
- ✿ 实验 2 分析成绩单
- ✿ 实验 3 文件读取和模式匹配
- ✿ 实验 4 读写基本类型数据
- ✿ 实验 5 对象的写入和读取
- ✿ 实验 6 使用 Scanner 解析文件
- ✿ 知识扩展——ZIP 文件的读取和制作

9.1 知识概述

当程序需要读取磁盘上的数据或将程序中的数据存储到磁盘时,就可以使用输入流和输出流,简称 I/O 流。I/O 流提供一条通道程序,可以使用这条通道读取"源"中的数据,或把数据送到"目的地"。I/O 流中的输入流的指向称为源,程序从指向源的输入流中读取源中的数据;输出流的指向称为目的地,程序通过向输出流中写入数据把信息送到目的地。

本章主要讲述下列流。

1. 文件字节流

FileInputStream 类是 InputStream 的子类,该类创建的对象称为文件字节输入流,用于按字节读取文件中的数据。FileInputStream 流顺序地读取文件,只要不关闭流,每次调用读取方法时就顺序地读取文件中其余的内容,直到文件的末尾或流被关闭。

FileOutputStream 类是 OutputStream 的子类,该类创建的对象称为文件字节输出流,用于输出流按字节将数据写入到文件中。FileOutStream 流顺序地写文件,只要不关闭流,每次调用写入方法就顺序地向文件写入内容,直到流被关闭。

2. 文件字符流

FileReader 类是 Reader 的子类,该类创建的对象称为文件字符输入流,用于输入流按字符读取文件中的数据。FileReader 流顺序地读取文件,只要不关闭流,每次调用读取方法时就顺序地读取文件中其余的内容,直到文件的末尾或流被关闭。

FileWriter 类是 Writer 的子类,该类创建的对象称为文件字符输出流,用于按字符将数据写入到文件中。FileWriter 流顺序地写文件,只要不关闭流,每次调用写入方法就顺序地向文件写入内容,直到流被关闭。

3. 缓冲流

BufferedReader 类创建的对象称为缓冲输入流,该输入流的指向必须是一个 Reader 流,称为 BufferedReader 流的底层流,负责将数据读入缓冲区。BufferedReader 流的源就是一个缓冲区,缓冲输入流,再从缓冲区中读取数据。

BufferedWriter 类创建的对象称为缓冲输出,可以将 BufferedWriter 流和 FileWriter 流连接在一起,然后使用 BufferedWriter 流将数据写到目的地。FileWriter 流称为 BufferedWriter 的底层流。BufferedWriter 流将数据写入缓冲区,底层流负责将数据写到最终的目的地。

4. 数组流

ByteArrayInputStream 类和 ByteArrayOutputStream 类创建的对象称为字节数组输入流和字节数组输出流,分别使用字节数组作为流的源和目的。与之对应的是字符数组类 CharArrayReader 和 CharArrayWriter,分别使用字符数组作为流的源和目的。

5. 字符串流

StringReader 类使用字符串作为流的源,StringWriter 类将内存作为流的目的地。

6. 数据流

DataInputStream 类和 DataOutputStream 类创建的对象称为数据输入流和数据输出流,它们可以以与机器无关的风格读取 Java 原始数据。

7．对象流

ObjectInputStream 类和 ObjectOutputStream 类创建的对象被称为对象输入流和对象输出流。对象输出流可以将一个对象写入输出流送往目的地,对象输入流可以从源中读取一个对象到程序中。当我们使用对象流写入或读入对象时,要保证对象是序列化的。这是为了保证能把对象写入到文件,并能再把对象正确读回到程序中。

一个类如果实现了 Serializable 接口,那么这个类创建的对象就是所谓序列化的对象。Serializable 接口中的方法对程序是不可见的,因此实现该接口的类不需要实现额外的方法。

8．随机读写流

RandomAccessFile 类创建的流的指向既可以作为源也可以作为目的地,称为随机读写流。对一个文件进行读或写操作时,可以创建一个指向该文件的 RandomAccessFile 流,这样既可以从这个流中读取文件的数据,也可以通过这个流写入数据到文件。

另外,本书主教材还介绍了与流有关的克隆对象、文件锁等知识。使用对象流很容易获取一个序列化对象的克隆,我们只需将该对象写入到对象输出流,然后用对象输入流读回的对象就是原对象的一个克隆。RondomAccessFile 类创建的流在读或写文件时可以使用文件锁,只要不解除该锁,其他线程无法操作被锁定的文件。

9.2 实验练习

9.2.1 文件加密

1．实验目的

本实验的目的是让学生掌握字符输入/输出流的用法。

2．实验要求

编写一个 Java 应用程序,将已存在的扩展名为 .txt 的文本文件加密后存入另一个文本文件中。

3．运行效果示例

运行效果如图 9-1 所示。

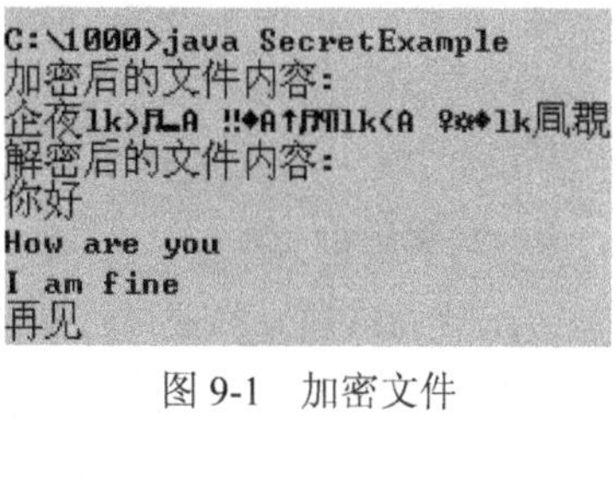

4．程序模板

按模板要求,将【代码】替换为程序代码。

图 9-1 加密文件

SecretExample.java

```
import java.io.*;
public class SecretExample{
    public static void main(String args[]){
        File fileOne=new File("hello.txt"), fileTwo=new File("hello.secret");
        char b[]=new char[100];
        try{
            FileReader in=【代码 1】            //创建指向 fileOne 的字符输入流
            FileWriter out=【代码 2】           //创建指向 fileTwo 的字符输出流
            int n=-1;
            while((n=in.read(b))!=-1){
```

```java
                    for(int i=0;i<n;i++){
                        b[i]=(char)(b[i]^'a');
                    }
                    【代码3】                          //out 将数组 b 的前 n 单元写到文件
                }
                out.close();
                in=【代码4】                          //创建指向 fileTwo 的字符输入流
                System.out.println("加密后的文件内容: ");
                while((n=in.read(b))!=-1){
                    String str=new String(b,0,n);
                    System.out.println(str);
                }
                in=new FileReader(fileTwo);
                System.out.println("解密后的文件内容: ");
                while((n=in.read(b))!=-1){
                    for(int i=0;i<n;i++){
                        b[i]=(char)(b[i]^'a');
                    }
                    System.out.printf(new String(b,0,n));
                }
                in.close();
            }
            catch(IOException e){
                System.out.println(e);
            }
        }
    }
```

5. 实验指导与检查

- 输入流 FileReader 调用 int read(char b[])方法从源中读取 b.length 个字符到字符数组 b 中，返回实际读取的字符数目。如果到达文件的末尾，则返回-1。
- FileWriter 调用 void write(char b[],int off,int len)方法把字符数组 b 中索引 off 处后的 len 个字符写入到输出流。
- 向实验指导教师演示程序的运行效果。

6. 实验报告

实验报告的格式如下（可要求学生填写并由实验指导教师签字）：

学号：_____ 班级：_____ 姓名：_____ 时间：_____

实 验 内 容	回 答	教 师 评 语
将上述程序中的字符流改成字节流		

9.2.2 分析成绩单

1. 实验目的

本实验的目的是让学生掌握字符输入流、输出流和缓冲流的用法。

2. 实验要求

现在有如下格式的成绩单（文本格式）score.txt：

　　姓名：张三，数学 72 分，物理 67 分，英语 70 分。
　　姓名：李四，数学 92 分，物理 98 分，英语 88 分。
　　姓名：周五，数学 68 分，物理 80 分，英语 77 分。

要求按行读取成绩单，并在该行的后面尾加上该同学的总成绩，再将该行写入到一个名为 socreAnalysis.txt 的文件中。

3. 运行效果示例

运行效果如图 9-2 所示。

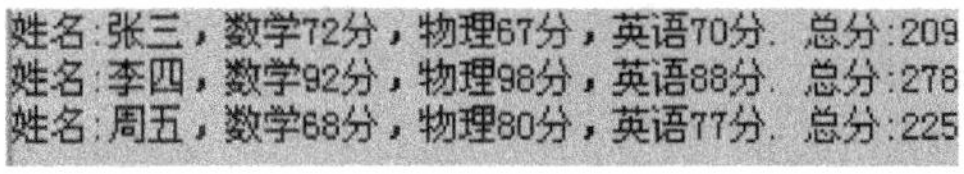

图 9-2　分析成绩单

4. 程序模板

按模板要求，将【代码】替换为程序代码。

AnalysisResult.java

```java
import java.io.*;
import java.io.*;
import java.util.*;
public class AnalysisResult {
    public static void main(String args[]){
        File fRead=new File("score.txt");
        File fWrite=new File("socreAnalysis.txt");
        try{
            Writer out=【代码1】              //以尾加方式创建指向文件 fWrite 的 out 流
            BufferedWriter bufferWrite=【代码2】   //创建指向 out 的 bufferWrite 流
            Reader in=【代码3】                //创建指向文件 fRead 的 in 流
            BufferedReader bufferRead=【代码4】  //创建指向 in 的 bufferRead 流
            String str=null;
            while((str=bufferRead.readLine())!=null){
                double totalScore=Fenxi.getTotalScore(str);
                str=str+ "总分: "+ totalScore;
                System.out.println(str);
                bufferWrite.write(str);
                bufferWrite.newLine();
            }
            bufferRead.close();
            bufferWrite.close();
        }
        catch(IOException e){
            System.out.println(e.toString());
        }
    }
}
```

Fenxi.java

```java
import java.util.*;
public class Fenxi{
```

```java
            public static double getTotalScore(String s)
                Scanner scanner=new Scanner(s);
                scanner.useDelimiter("[^0123456789.]+");
                double totalScore=0;
                while(scanner.hasNext()){
                    try{
                        double score=scanner.nextDouble();
                        totalScore=totalScore+ score;
                    }
                    catch(InputMismatchException exp){
                        String t=scanner.next();
                    }
                }
                return totalScore;
            }
        }
```

5. 实验指导与检查

- BufferedReader 类创建的对象称为缓冲输入流，该输入流的指向必须是一个 Reader 流，称为 BufferedReader 流的底层流，负责将数据读入缓冲区。BufferedReader 流的源就是这个缓冲区，缓冲输入流再从缓冲区中读取数据。
- BufferedWriter 类创建的对象称为缓冲输出，可以将 BufferedWriter 流和 FileWriter 流连接在一起，然后使用 BufferedWriter 流将数据写到目的地。
- BufferedReader 对象调用 readLine()方法可读取文件的一行。
- BufferedWriter 对象调用 newLine()方法可向文件写入回行。
- 向实验指导教师演示程序的运行效果。

6. 实验报告

实验报告的格式如下（可要求学生填写并由实验指导教师签字）：

学号：_____ 班级：_____ 姓名：_____ 时间：_____

实 验 内 容	回　　答	教 师 评 语
改进程序，使得能统计出每个学生的平均成绩		

9.2.3 文件读取和模式匹配

1. 实验目的

本实验的目的是让学生掌握使用流读取文件的内容，并将内容写入到内存，然后使用模式匹配技术查找文件内容和指定模式匹配的字符序列。

2. 实验要求

编写一个 Java 应用程序，将一个文件的内容读入到程序中，然后输出该文件使用了哪些单词。

3. 运行效果示例

运行效果如图 9-3 所示。

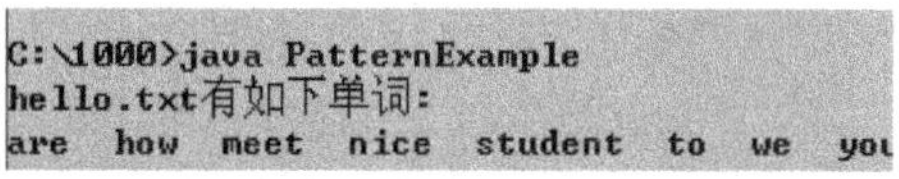

4. 程序模板

按模板要求，将【代码】替换为程序代码。

图 9-3　文件读取与模式匹配

PatternExample.java

```
import java.io.*;
import java.util.*;
import java.util.regex.*;
public class PatternExample{
    public static void main(String args[]){
        File file=new File("hello.txt");
        try{
            FileReader firstIn=【代码1】             //创建指向file的字符输入流
            BufferedReader secondIn=【代码2】        //创建指向firstIn的缓冲流
            StringWriter out=【代码3】               //创建指向内存的字符串输出流
            String s=null;
            while((s=secondIn.readLine())!=null){
                【代码4】                             //out将s写入内存
            }
            String content=【代码5】                 //out获取曾写入到内存中的全部字符
            firstIn.close();
            secondIn.close();
            out.close();
            Pattern p;
            Matcher m;
            p=Pattern.compile("[a-z[A-Z]]+");
            m=p.matcher(content);
            TreeSet<String> tree=new TreeSet<String>();
            while(m.find()){
                String str=m.group();
                tree.add(str);
            }
            Iterator<String> iter=tree.iterator();
            System.out.println(file.getName()+"使用了如下单词：");
            while(iter.hasNext()){
                String item=iter.next();
                System.out.print (item+"  ");
            }
        }
        catch(IOException e){
            System.out.println(e);
        }
    }
}
```

5．实验指导与检查

⊙ StringWriter 将内存作为流的目的地，使用 StringWriter()方法构造字符串输出流对象，指向一个默认大小的缓冲区。输出流向缓冲区写入的字符所占内存的总量大于缓冲区时，缓冲区的容量会自动增加。
⊙ "[a-z[A-Z]]+"模式可以匹配英文单词。
⊙ 向实验指导教师演示程序的运行效果。

6．实验报告

实验报告的格式如下（可要求学生填写并由实验指导教师签字）：

学号：_____　班级：_____　姓名：_____　时间：_____

实 验 内 容	回　答	教 师 评 语
编写一个应用程序，输出该文件使用了哪些标点符号、文件内容中出现了几次以数字为前缀的单词		

9.2.4 读/写基本类型数据

1．实验目的

本实验的目的是让学生掌握数据流的使用方法。

2．实验要求

编写一个 Java 应用程序，将若干基本数据写入到一个文件，再按顺序读出。

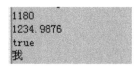

图 9-4　读写基本类型数据

3．运行效果示例

运行效果如图 9-4 所示。

4．程序模板

按模板要求，将【代码】替换为程序代码。

NumberExample.java

```
import java.io.*;
public class NumberExample{
    public static void main(String args[]){
        int a=1180;
        double d=1234.9876;
        boolean boo=true;
        char c='我';
        File f=new File("jerry.dat");
        try{
            FileOutputStream fos=new FileOutputStream(f);
            DataOutputStream out_data=new DataOutputStream(fos);
            【代码1】                    //out_data 将数据 a 写入到文件中
            【代码2】                    //out_data 将数据 d 写入到文件中
            【代码3】                    //out_data 将数据 boo 写入到文件中
            【代码4】                    //out_data 将数据 c 写入到文件中
```

```
            }
            catch(IOException e){ }
            try{
                FileInputStream fis=new FileInputStream(f);
                DataInputStream in_data=new DataInputStream(fis);
                System.out.println(in_data.readInt());
                System.out.println(in_data.readDouble());
                System.out.println(in_data.readBoolean());
                System.out.print(in_data.readChar());
            }
            catch(IOException e){ }
        }
    }
```

5．实验指导与检查

- DataInputStream 类和 DataOutputStream 类创建的对象称为数据输入流和数据输出流，它们允许程序按照与机器无关的风格读取 Java 原始数据。也就是说，当读取一个数值时，不必关心这个数值应当是多少字节。
- 向指导教师演示程序的运行效果。

6．实验报告

实验报告的格式如下（可要求学生填写并由实验指导教师签字）：

学号：_____ 班级：_____ 姓名：_____ 时间：_____

实验内容	回　　答	教师评语
在上述程序中添加用 writeUTF()方法，将字符串"你好，很高兴认识你"写入到文件中，再读回到程序中的代码		

9.2.5　对象的写入和读取

1．实验目的

本实验的目的是让学生掌握对象流的使用。

2．实验要求

编写一个 Java 应用程序，将一个 Calendar 对象写入文件，然后顺序读出该对象，并验证读出的对象是原始对象的克隆。

3．运行效果示例

运行效果如图 9-5 所示。

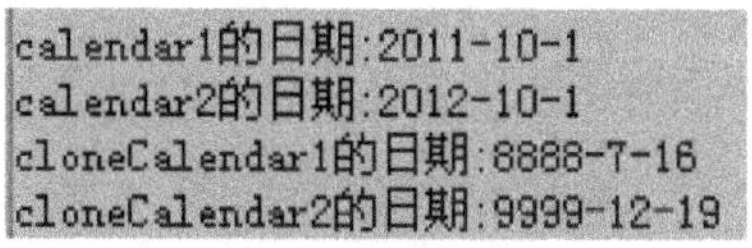

```
calendar1的日期:2011-10-1
calendar2的日期:2012-10-1
cloneCalendar1的日期:8888-7-16
cloneCalendar2的日期:9999-12-19
```

图 9-5　读、写对象

4．程序模板

按模板要求，将【代码】替换为程序代码。

ObjectExample.java

```
import java.util.*;
import java.io.*;
```

```java
public class ObjectExample{
    public static void main(String args[]){
        Calendar calendar1=Calendar.getInstance();
        Calendar calendar2=Calendar.getInstance();
        calendar1.set(2011,9,1);      //将日历时间设置为 2011 年 10 月 1 日，注意 9 表示十月
        calendar2.set(2012,9,1);
        try{
            File f=new File("a.txt");
            FileOutputStream fileOut=new FileOutputStream(f);
            ObjectOutputStream objectOut=new ObjectOutputStream(fileOut);
            【代码 1】                              //objectOut 写对象 calendar1 到文件
            【代码 2】                              //objectOut 写对象 calendar2 到文件
            FileInputStream fileIn=new FileInputStream(f);
            ObjectInputStream objectIn=new ObjectInputStream(fileIn);
            Calendar cloneCalendar1= 【代码 3】     //objectOut 读出对象
            Calendar cloneCalendar2= 【代码 4】     //objectOut 读出对象
            cloneCalendar1.set(8888,6,16);
            cloneCalendar2.set(9999,11,19);
            showCalendar(calendar1,"calendar1");
            showCalendar(calendar2,"calendar2");
            showCalendar(cloneCalendar1,"cloneCalendar1");
            showCalendar(cloneCalendar2,"cloneCalendar2");
        }
        catch(Exception event){
            System.out.println(event);
        }
    }
    public static void showCalendar(Calendar calendar,String name){
        int 年=calendar.get(Calendar.YEAR),
            月=calendar.get(Calendar.MONTH)+ 1,
            日=calendar.get(Calendar.DAY_OF_MONTH);
        System.out.printf("\n%s 的日期: %d-%d-%d",name,年,月,日);
    }
}
```

5. 实验指导与检查

- 使用对象流写入或读入对象时，要保证对象是序列化的。这是为了保证能把对象写入到文件中，并能再把对象正确读回到程序中。Java 提供的绝大多数对象都是序列化的。一个类如果实现了 Serializable 接口，那么这个类创建的对象就是序列化的对象，Calender 类实现了 Serializable 接口，其中的方法对程序是不可见的，因此实现该接口的类不需要实现额外的方法。当把一个序列化的对象写入到对象输出流时，JVM 就会实现 Serializable 接口中的方法，按照一定格式的文本，将对象写入到目的地。
- 使用对象流很容易获取一个序列化对象的克隆。我们只需将该对象写入到对象输出流，用对象输入流读回的对象就是原对象的一个克隆。
- 向实验指导教师演示程序的运行效果。

6. 实验报告

实验报告的格式如下（可要求学生填写并由实验指导教师签字）：

学号：_____ 班级：_____ 姓名：_____ 时间：_____

实 验 内 容	回　　答	教 师 评 语
编写一个 Student 类，将该类创建的若干个对象写入到文件，然后读出这些对象		

9.2.6 使用 Scanner 解析文件

1. 实验目的

掌握使用 Scanner 类解析文件的方法。

2. 实验要求

使用 Scanner 类和正则表达式统计一篇英文中的单词，要求如下：① 一共出现了多少个单词；② 有多少个互不相同的单词；③ 按单词出现频率大小输出单词。

3. 运行效果示例

运行效果如图 9-6 所示。

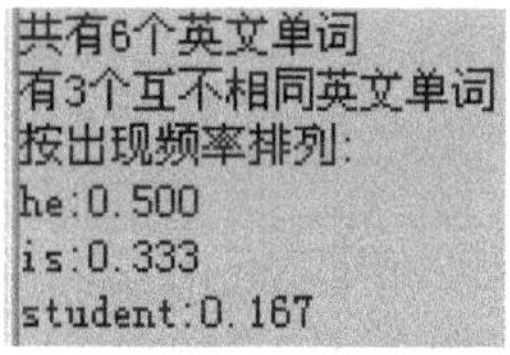

4. 程序模板

按模板要求，将【代码】替换为程序代码。

图 9-6　统计单词

WordStatistic.java

```
import java.io.*;
import java.util.*;
public class WordStatistic{
    Vector<String> allWord,noSameWord;
    File file=new File("english.txt");
    Scanner sc=null;
    String regex;
    WordStatistic(){
        allWord=new Vector<String>();
        noSameWord=new Vector<String>();
            //regex是由空格、数字和符号(!"#$%&'()*+,-./:;<=>?@[\]^_`{|}~)组成的正则表达式
        regex= "[\\s\\d\\p{Punct}]+";
        try{
            sc=【代码1】         //创建指向 file 的 sc
            【代码2】             //sc 调用 useDelimiter(String regex)方法，向参数传递 regex
        }
        catch(IOException exp){
            System.out.println(exp.toString());
        }
    }
    void setFileName(String name){
        file=new File(name);
        try{
```

```java
            sc=new Scanner(file);
            sc.useDelimiter(regex);
        }
        catch(IOException exp){
            System.out.println(exp.toString());
        }
    }
    public void wordStatistic(){
        try{
            while(sc.hasNext()){
                String word = sc.next();
                allWord.add(word);
                if(!noSameWord.contains(word)){
                    noSameWord.add(word);
                }
            }
        }
        catch(Exception e){ }
    }
    public Vector<String> getAllWord(){
        return allWord;
    }
    public Vector<String> getNoSameWord(){
        return noSameWord;
    }
}
```

OutputWordMess.java

```java
import java.util.*;
public class OutputWordMess{
    public static void main(String args[]){
        Vector<String> allWord,noSameWord;
        WordStatistic statistic =new WordStatistic();
        statistic.setFileName("english.txt");
        【代码3】                                      //statistic 调用 wordStatistic()方法
        allWord=statistic.getAllWord();
        noSameWord=statistic.getNoSameWord();
        System.out.println("共有"+ allWord.size()+ "个英文单词");
        System.out.println("有"+ noSameWord.size()+ "个互不相同英文单词");
        System.out.println("按出现频率排列: ");
        int count[]=new int[noSameWord.size()];
        for(int i=0;i<noSameWord.size();i++){
            String s1 = noSameWord.elementAt(i);
            for(int j=0;j<allWord.size();j++){
                String s2=allWord.elementAt(j);
                if(s1.equals(s2)){
                    count[i]++;
                }
```

```
                }
            }
            for(int m=0;m<noSameWord.size();m++){
                for(int n=m+1;n<noSameWord.size();n++){
                    if(count[n]>count[m]){
                        String temp=noSameWord.elementAt(m);
                        noSameWord.setElementAt(noSameWord.elementAt(n),m);
                        noSameWord.setElementAt(temp,n);
                        int t=count[m];
                        count[m]=count[n];
                        count[n]=t;
                    }
                }
            }
            for(int m=0;m<noSameWord.size();m++){
                double frequency=(1.0*count[m])/allWord.size();
                System.out.printf("%s: %-7.3f",noSameWord.elementAt(m),frequency);
            }
        }
    }
```

5. 实验指导与检查

- java.util 包中的 Vector 类负责创建一个向量对象。如果已经会使用数组，那么很容易就会使用向量。创建一个向量时不用像数组那样必须要给出数组的大小。例如，向量创建后，对于 Vector<String> a=new Vector<String>()方法，a 可以使用 add(String o)方法把 String 对象添加到向量的末尾，向量的大小会自动增加；a 可以使用 elementAt(int index)方法获取指定索引处的向量的元素（索引初始位置是 0）；a 可以使用 size()方法获取向量所含有的元素的个数。

- 如果 Scanner 对象不使用 useDelimiter 设置，而是使用正则表达式作为分隔标记，那么 Scanner 对象使用空格作为分隔标记。

- 向实验指导教师演示程序的运行效果。

6. 实验报告

实验报告的格式如下（可要求学生填写并由实验指导教师签字）：

学号：_____ 班级：_____ 姓名：_____ 时间：_____

实 验 内 容	回　　答	教师评语
按字典序输出全部不相同的单词		

9.3　知识扩展——ZIP 文件的读取和制作

1. ZipInputStream 类

ZIP 文件是一种流行的档案文件，可以将若干个文件压缩到一个 ZIP 文件中。

使用 ZipInputStream 类创建的输入流对象可以读取压缩到 ZIP 文件中的各文件（解压）。假设要解压一个名字为 book.zip 的文件，需要首先使用 ZipInputStream 的构造方法 public ZipInputStream(InputStream in)创建一个对象 in，如

      ZipInputStream in=new ZipInputStream(new FileInputStream("book.zip"));

然后让 **ZipInputStream** 对象 in 找到 book.zip 中的下一个文件，如

      ZipEntry zipEntry=in.getNextEntry();

那么，in 调用 read()方法可以读取找到的该文件。

在下面的 ReadZipFile.java 文件中，读取含有两个文件的 book.zip，并将这两个文件重新存放到当前目录的 book 文件夹中，即将 book.zip 的内容解压到 book 文件夹中。

```
ReadZipFile.java
import java.io.*;
import java.util.zip.*;
public class ReadZipFile{
    public static void main(String args[]){
        File f=new File("book.zip");
        File dir=new File("Book");
        byte b[]=new byte[100];
        dir.mkdir();
        try{
            ZipInputStream in=new ZipInputStream(new FileInputStream(f));
            ZipEntry zipEntry=null;
            while((zipEntry=in.getNextEntry())!=null){
                File file=new File(dir,zipEntry.getName());
                FileOutputStream out=new FileOutputStream(file);
                int n=-1;
                while((n=in.read(b,0,100))!=-1){
                    String str=new String(b,0,n);
                    System.out.println(str);
                    out.write(b,0,n);
                }
                out.close();
            }
            in.close();
        }
        catch(IOException ee){
            System.out.println(ee);
        }
    }
}
```

2．ZipOutputStream 类

使用 ZipOutputStream 类可以制作 ZIP 文件，即将若干个文件压缩到一个 ZIP 文件中。

首先，用 ZipOutputStream 类的构造方法 public ZipOutputStream(OutputStream out)创建一个对象，如

ZipOutputStream out=new ZipOutputStream(new FileOutputStream("first.zip"));

再让 ZipOutputStream 对象 out 调用 public void putNextEntry(ZipEntry e)方法，确定向 ZIP 文件写入的下一个文件的位置，如

out.putNextEntry(new ZipEntry("A.txt"));

然后，out 对象调用 write()方法，将文件 A.txt 写入到 ZIP 文件中（压缩过程）。

下面的 WiteZipFile.java 将 3 个文件压缩到一个 ZIP 文件中。

WiteZipFile.java

```java
import java.io.*;
import java.util.zip.*;
public class WiteZipFile {
    public static void main(String args[]) {
        File f=new File("first.zip");
        try {
            ZipOutputStream out=new ZipOutputStream(new FileOutputStream(f));
            out.putNextEntry(new ZipEntry("Example.java"));
            FileInputStream reader=new FileInputStream("Example.java");
            byte b[]=new byte[100];
            int n=-1;
            while((n=reader.read(b,0,100))!=-1) {
                out.write(b,0,n);                    //写数据到 first.zip 文件中
            }
            out.putNextEntry(new ZipEntry("a.txt"));
            reader=new FileInputStream("a.txt");
            n=-1;
            while((n=reader.read(b,0,100))!=-1) {
                out.write(b,0,n);                    //写数据到 first.zip 文件中
            }
            out.putNextEntry(new ZipEntry("hello.txt"));
            reader=new FileInputStream("hello.txt");
            n=-1;
            while((n=reader.read(b,0,100))!=-1) {
                out.write(b,0,n);                    //写数据到 first.zip 文件中
            }
            out.close();
        }
        catch(IOException ee) {
            System.out.println(ee);
        }
    }
}
```

第 10 章

图形用户界面设计

本章导读

- ✿ 知识概述
- ✿ 实验 1 布局与日历
- ✿ 实验 2 猜数字游戏
- ✿ 实验 3 算术测试
- ✿ 实验 4 单词统计与排序
- ✿ 实验 5 华容道游戏
- ✿ 实验 6 字体对话框
- ✿ 知识扩展——计时器

10.1 知识概述

通过图形用户界面（Graphics User Interface，GUI），用户与程序之间可以方便地进行交互。Java 包含了许多支持 GUI 设计的类，如按钮、菜单、列表、文本框等组件类，同时包含窗口、面板等容器类。学习组件除了了解组件的属性和功能外，一个更重要的方面是学习怎样处理组件上发生的界面事件。在学习处理事件时，必须很好地掌握事件源、监视器、处理事件的接口这三个概念。

① 事件源。能够产生事件的对象都可以成为事件源，如文本框、按钮、下拉式列表等。也就是说，事件源必须是一个对象，而且这个对象必须是 Java 认为能够发生事件的对象。

② 监视器。我们需要一个对象对事件源进行监视，以便对发生的事件做出处理，事件源通过调用相应的方法将某个对象作为自己的监视器。

③ 处理事件的接口。监视器负责处理事件源发生的事件。Java 语言使用了接口回调技术来设计它的处理事件模式。事件源增加监视的方法 addXXXListener(XXXListener listener)中的参数是一个接口，listener 可以引用任何实现了该接口的类所创建的对象，当事件源发生事件时，接口 listener 立刻回调被类实现的接口中的某个方法。

10.2 实验练习

10.2.1 布局与日历

1．实验目的

本实验的目的是学习使用布局类。

2．实验要求

编写一个应用程序，有一个窗口，该窗口的布局为 BorderLayout 布局。窗口的中心添加一个 JPanel 容器 pCenter，pCenter 的布局是 7 行 7 列的 GriderLayout 布局，放置 49 个标签，用来显示日历。窗口的北面添加一个 JPanel 容器 pNorth，其布局是 FlowLayout 布局，pNorth 放置两个按钮：nextMonth 和 previousMonth，单击 nextMonth 按钮，可以显示当前月的下一月的日历；单击 previousMonth 按钮，可以显示当前月的上一月的日历。窗口的南面添加一个 JPanel 容器 pSouth，其布局是 FlowLayout 布局，pSouth 中放置一个标签用来显示一些信息。

3．运行效果示例

运行效果如图 10-1 所示。

4．程序模板

按模板要求，将【代码】替换为程序代码。

CalendarMainClass.java

```
public class CalendarMainClass{
    public static void main(String args[]){
        CalendarFrame frame=new CalendarFrame();
        frame.setBounds(100,100,360,300);
```

图 10-1　布局与日历

```java
            frame.setVisible(true);
            frame.setYearAndMonth(2015,5);
        }
    }
```

CalendarBean.java

```java
    import java.util.Calendar;
    public class CalendarBean{
        String day[];
        int year=2005,month=0;
        public void setYear(int year){
            this.year=year;
        }
        public int getYear(){
            return year;
        }
        public void setMonth(int month){
            this.month=month;
        }
        public int getMonth(){
            return month;
        }
        public String[] getCalendar(){
            String a[]=new String[42];
            Calendar 日历=Calendar.getInstance();
            日历.set(year,month-1,1);
            int 星期几=日历.get(Calendar.DAY_OF_WEEK)-1;
            int day=0;
            if(month==1 || month==3 || month==5 || month==7 || month==8 || month==10 || month==12){
                day=31;
            }
            if(month==4 || month==6 || month==9 || month==11){
                day=30;
            }
            if(month==2){
                if(((year%4==0) && (year%100!=0)) || (year%400==0)){
                    day=29;
                }
                else{
                    day=28;
                }
            }
            for(int i=星期几,n=1;i<星期几+day;i++){
                a[i]=String.valueOf(n) ;
                n++;
            }
            return a;
        }
    }
```

CalendarFrame.java
```
import java.util.*;
import java.awt.*;
import java.awt.event.*;
import javax.swing.*;
import javax.swing.border.*;
public class CalendarFrame extends JFrame implements ActionListener{
    JLabel labelDay[]=new JLabel[42];
    JButton titleName[]=new JButton[7];
    String name[]={"日","一","二","三","四","五","六"};
    JButton nextMonth,previousMonth;
    CalendarBean calendar;
    JLabel showMessage=new JLabel("",JLabel.CENTER);
    int year=2011,month=2;
    public CalendarFrame(){
        JPanel pCenter=new JPanel();
        【代码1】                    //将 pCenter 的布局设置为 7 行 7 列的 GridLayout 布局
        for(int i=0;i<7;i++){
            titleName[i]=new JButton(name[i]);
            titleName[i].setBorder(new SoftBevelBorder(BevelBorder.RAISED));
            pCenter.add(titleName[i]);
        }
        for(int i=0;i<42;i++){
            labelDay[i]=new JLabel("",JLabel.CENTER);
            labelDay[i].setBorder(new SoftBevelBorder(BevelBorder.LOWERED));
            【代码2】                    //在 pCenter 中添加组件 labelDay[i]
        }
        calendar=new CalendarBean();
        nextMonth=new JButton("下月");
        previousMonth=new JButton("上月");
        nextMonth.addActionListener(this);
        previousMonth.addActionListener(this);
        JPanel pNorth=new JPanel(), pSouth=new JPanel();
        pNorth.add(previousMonth);
        pNorth.add(nextMonth);
        pSouth.add(showMessage);
        add(pCenter,BorderLayout.CENTER);
        【代码3】                    //在窗口中添加 pNorth，在北区域
        【代码4】                    //在窗口中添加 pSouth，在南区域
        setYearAndMonth(year,month);
        setDefaultCloseOperation(DISPOSE_ON_CLOSE);
    }
    public void setYearAndMonth(int y,int m){
        calendar.setYear(y);
        calendar.setMonth(m);
        String day[]=calendar.getCalendar();
        for(int i=0;i<42;i++){
            labelDay[i].setText(day[i]);
```

```java
            }
            showMessage.setText("日历: "+ calendar.getYear()+"年"+ calendar.getMonth()+ "月");
        }
        public void actionPerformed(ActionEvent e){
            if(e.getSource()==nextMonth){
                month=month+1;
                if(month>12){
                    month=1;
                }
                calendar.setMonth(month);
                String day[]=calendar.getCalendar();
                for(int i=0;i<42;i++){
                    labelDay[i].setText(day[i]);
                }
            }
            else if(e.getSource()==previousMonth){
                month=month-1;
                if(month<1){
                    month=12;
                }
                calendar.setMonth(month);
                String day[]=calendar.getCalendar();
                for(int i=0;i<42;i++){
                    labelDay[i].setText(day[i]);
                }
                showMessage.setText("日历:"+ calendar.getYear()+ "年"+ calendar.getMonth()+ "月");
            }
        }
```

5. 实验指导与检查

- BorderLayout 是一种简单的布局策略，如果一个容器使用这种布局，那么容器空间简单地划分为东、西、南、北、中五个区域，中间的区域最大。每加入一个组件都应该指明把这个组件添加在哪个区域中。区域由 BorderLayout 中的静态常量 CENTER、NORTH、SOUTH、WEST、EAST 表示。
- GridLayout 是使用较多的布局编辑器，其基本布局策略是把容器划分成若干行×若干列的网格区域，组件就位于这些划分出来的小格中。GridLayout 比较灵活，划分多少网格由程序自由控制，而且组件定位也比较精确。
- 向实验指导教师演示程序的运行效果。

6. 实验报告

实验报告的格式如下（可要求学生填写并由实验指导教师签字）：

学号：_____ 班级：_____ 姓名：_____ 时间：_____

实 验 内 容	回 答	教师评语
请在 CalendarFrame 类中增加一个 JTextField 文本框,用户可以通过在文本框中输入年份来修改 calendar 对象的 int 成员 year		

10.2.2 猜数字游戏

1．实验目的

本实验的目的是让学生学习处理 ActionEvent 事件的方法。

2．实验要求

按钮被单击，可以发生 ActionEvent 事件，当按钮获得监视器之后，单击它，就发生 ActionEven 事件，即 java.awt.envent 包中的 ActionEvent 类自动创建了一个事件对象。要求界面中有两个按钮 buttonGetNumber 和 buttonEnter，用户单击 buttonGetNumber 按钮可以获得一个随机数，然后在一个文本框中输入猜测，再单击 buttonEnter 按钮，程序根据用户的猜测给出提示信息。

3．运行效果示例

运行效果如图 10-2 所示。

4．程序模板

按模板要求，将【代码】替换为程序代码。

图 10-2　猜数字

GuessExample.java

```
import java.awt.*;
import java.awt.event.*;
import javax.swing.*;
import javax.swing.border.*;
class WindowButton extends JFrame implements ActionListener{
    int number;
    JTextField inputNumber;
    JLabel feedBack;
    JButton buttonGetNumber,buttonEnter;
    WindowButton(String s){
        super(s);
        buttonGetNumber=new JButton("得到一个随机数");
        feedBack=new JLabel("无反馈信息",JLabel.CENTER);
        feedBack.setBackground(Color.green);
        inputNumber=new JTextField("0",5);
        buttonEnter=new JButton("确定");
        【代码1】    //按钮 buttonEnter 增加 ActionEvent 事件监视器，监视器为当前窗口
        【代码2】    //按钮 buttonGetNumber 增加 ActionEvent 事件监视器，监视器为当前窗口
        Box boxH1=Box.createHorizontalBox();
        boxH1.add(new JLabel("获取 1～100 之间的随机数："));
        boxH1.add(buttonGetNumber);
        Box boxH2=Box.createHorizontalBox();
        boxH2.add(new JLabel("输入您的猜测："));
        boxH2.add(inputNumber);
        Box boxH3=Box.createHorizontalBox();
        boxH3.add(new JLabel("单击确定按钮："));
        boxH3.add(buttonEnter);
```

```java
        Box boxH4=Box.createHorizontalBox();
        boxH4.add(new JLabel("反馈信息: "));
        boxH4.add(feedBack);
        Box baseBox=Box.createVerticalBox();
        baseBox.add(boxH1);
        baseBox.add(boxH2);
        baseBox.add(boxH3);
        baseBox.add(boxH4);
        Container con=getContentPane();
        con.setLayout(new FlowLayout());
        con.add(baseBox);
        con.validate();
        setBounds(120,125,270,200);
        setVisible(true);
        setDefaultCloseOperation(JFrame.EXIT_ON_CLOSE);
        setBounds(100,100,150,150);
        setVisible(true);
        validate();
    }
    public void actionPerformed(ActionEvent e){
        if( 【代码3】 ){                      //判断事件源是否是 buttonGetNumber
            number=(int)(Math.random()*100)+ 1;
        }
        else if( 【代码4】 ){                 //判断事件源是否是 buttonEnter
            int guess=0;
            try{
                guess=Integer.parseInt(inputNumber.getText());
                if(guess==number){
                    feedBack.setText("猜对了! ");
                }
                else if(guess>number){
                    feedBack.setText("猜大了! ");
                    InputNumber.setText(null);
                }
                else if(guess<number){
                    feedBack.setText("猜小了! ");
                    inputNumber.setText(null);
                }
            }
            catch(NumberFormatException event){
                feedBack.setText("请输入数字字符");
            }
        }
    }
}
```

```
class GuessExample{
    public static void main(String args[]){
        new WindowButton("猜数字小游戏");
    }
}
```

5．实验指导与检查

- JButton 对象可以发生 ActionEvent 类型事件。为了能监视到这种类型的事件，事件源必须调用 addActionListener()方法获得监视器。创建监视器的类必须实现接口 ActionListener。
- 向实验指导教师演示程序的运行效果。

6．实验报告

实验报告的格式如下（可要求学生填写并由实验指导教师签字）：

学号：_____   班级：_____   姓名：_____   时间：_____

实 验 内 容	回　　答	教师评语
给上述程序增加记录猜测次数的功能，每次反馈这是第几次猜测，成功时反馈一共猜了几次		

10.2.3　算术测试

1．实验目的

本实验的目的是学习处理 ActionEvent 事件的方法。

2．实验要求

编写一个算术测试小软件，用来训练小学生的算术能力。程序由 3 个类组成：Teacher 类的对象充当监视器，负责给出算术题目，并判断回答者的答案是否正确；ComputerFrame 类的对象负责为算术题目提供视图，如用户可以通过 ComputerFrame 类的对象提供的 GUI 界面看到题目，并通过该 GUI 界面给出题目的答案；MailClass 类是软件的主类。

3．运行效果示例

运行效果如图 10-3 所示。

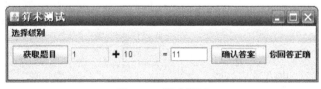

图 10-3　算术测试

4．程序模板

按模板要求，将【代码】替换为程序代码。

MainClass.java

```
public class MainClass {
    public static void main(String args[]) {
        ComputerFrame frame;
```

```
            frame=new ComputerFrame();
            frame.setTitle("算术测试");
            frame.setBounds(100,100,650,180);
       }
   }
```

ComputerFrame.java

```
       import java.awt.*;
       import java.awt.event.*;
       import javax.swing.*;
       public class ComputerFrame extends JFrame {
           JMenuBar menubar;
           JMenu choiceGrade;                                     //选择级别的菜单
           JMenuItem   grade1,grade2;
           JTextField textOne,textTwo,textResult;
           JButton getProblem,giveAnwser;
           JLabel operatorLabel,message;
           Teacher teacherZhang;
           ComputerFrame(){
              teacherZhang=new Teacher();
              teacherZhang.setMaxInteger(20);
              setLayout(new FlowLayout());
              menubar = new JMenuBar();
              choiceGrade = new JMenu("选择级别");
              grade1 = new JMenuItem("幼儿级别");
              grade2 = new JMenuItem("儿童级别");
              grade1.addActionListener(new ActionListener(){
                                  public void actionPerformed(ActionEvent e){
                                      teacherZhang.setMaxInteger(10);
                                  }
                              });
              grade2.addActionListener(new ActionListener(){
                                  public void actionPerformed(ActionEvent e){
                                      teacherZhang.setMaxInteger(50);
                                  }
                              });
              choiceGrade.add(grade1);
              choiceGrade.add(grade2);
              menubar.add(choiceGrade);
              setJMenuBar(menubar);
              【代码1】                                    //创建textOne，其可见字符数是5
              textTwo=new JTextField(5);
              textResult=new JTextField(5);
              operatorLabel=new JLabel("+ ");
              operatorLabel.setFont(new Font("Arial",Font.BOLD,20));
              message=new JLabel("你还没有回答呢");
```

```java
                    getProblem=new JButton("获取题目");
                    giveAnwser=new JButton("确认答案");
                    add(getProblem);
                    add(textOne);
                    add(operatorLabel);
                    add(textTwo);
                    add(new JLabel("="));
                    add(textResult);
                    add(giveAnwser);
                    add(message);
                    textResult.requestFocus();
                    textOne.setEditable(false);
                    textTwo.setEditable(false);
                    getProblem.setActionCommand("getProblem");
                    textResult.setActionCommand("answer");
                    giveAnwser.setActionCommand("answer");
                    teacherZhang.setJTextField(textOne,textTwo,textResult);
                    teacherZhang.setJLabel(operatorLabel,message);
                【代码2】           //将 teacherZhang 注册为 getProblem 的 ActionEvent 事件监视器
                【代码3】           //将 teacherZhang 注册为 giveAnwser 的 ActionEvent 事件监视器
                【代码4】           //将 teacherZhang 注册为 textResult 的 ActionEvent 事件监视器
                    setVisible(true);
                    validate();
                    setDefaultCloseOperation(DISPOSE_ON_CLOSE);
             }
        }
```

Teacher.java

```java
        import java.util.Random;
        import java.awt.event.*;
        import javax.swing.*;
        public class Teacher implements ActionListener {
            int numberOne,numberTwo;
            String operator="";
            boolean isRight;
            Random random;                              //用于给出随机数
            int maxInteger;                             //题目中最大的整数
            JTextField textOne,textTwo,textResult;
            JLabel operatorLabel,message;
            Teacher(){
                random=new Random();
            }
            public void setMaxInteger(int n) {
                maxInteger=n;
            }
            public void actionPerformed(ActionEvent e) {
```

```java
            String str = e.getActionCommand();
            if(str.equals("getProblem")){
                numberOne = random.nextInt(maxInteger)+ 1;    //1~maxInteger 之间的随机数
                numberTwo=random.nextInt(maxInteger)+ 1;
                double d=Math.random();                        //获取(0,1)之间的随机数
                if(d>=0.5){
                    operator="+ ";
                }
                else{
                    operator="-";
                }
                textOne.setText(""+ numberOne);
                textTwo.setText(""+ numberTwo);
                operatorLabel.setText(operator);
                message.setText("请回答");
                textResult.setText(null);
            }
            else if(str.equals("answer")) {
                String answer=textResult.getText();
                try{
                    int result=Integer.parseInt(answer);
                    if(operator.equals("+ ")){
                        if(result==numberOne+ numberTwo) {
                            message.setText("你回答正确");
                        }
                        else{
                            message.setText("你回答错误");
                        }
                    }
                    else if(operator.equals("-")){
                        if(result==numberOne-numberTwo) {
                            message.setText("你回答正确");
                        }
                        else{
                            message.setText("你回答错误");
                        }
                    }
                }
                catch(NumberFormatException ex) {
                    message.setText("请输入数字字符");
                }
            }
        }
        public void setJTextField(JTextField ... t) {
            textOne=t[0];
```

```
            textTwo=t[1];
            textResult=t[2];
        }
        public void setJLabel(JLabel ...label) {
            operatorLabel=label[0];
            message=label[1];
        }
    }
```

5．实验指导与检查

- 需要将实验中的 3 个 Java 文件保存在同一文件中，分别编译或只编译主类 MainClass，然后运行主类即可。
- JButton 对象可触发 ActionEvent 事件。为了能监视到此类型事件，事件源必须使用 addActionListener()方法获得监视器，创建监视器的类必须实现接口 ActionListener。
- 向实验指导教师演示程序的运行效果。

6．实验报告

实验报告的格式如下（可要求学生填写并由实验指导教师签字）：

学号：_____ 班级：_____ 姓名：_____ 时间：_____

实 验 内 容	回 答	教 师 评 语
模仿本实验代码，再增加"小学生"级别		

10.2.4 单词统计和排序

1．实验目的

本实验的目的是让学生学习使用 JTextArea 和处理 DucumentEvent 事件的方法。

2．实验要求

程序中有两个文本区，当用户在一个文本区中输入一篇英文短文时，另一个文本区将英文短文中使用的单词按字典序排列，并显示每个单词的使用频率。也就是说，随着输入的变化，另一个文本区不断更新显示的内容。

3．运行效果示例

运行效果如图 10-4 所示。

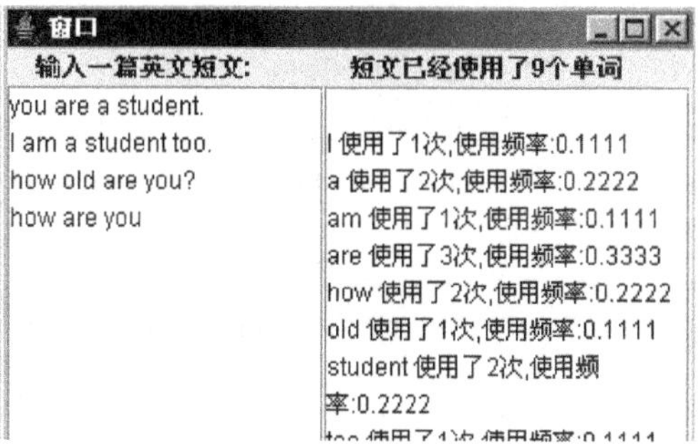

4．程序模板

按模板要求，将【代码】替换为程序代码。

SortExample.java

```
    import javax.swing.*;
    import java.awt.event.*;
    import java.awt.*;
    import java.util.regex.*;
    import java.util.*;
```

图 10-4 单词统计和排序

```java
import javax.swing.event.*;
import javax.swing.text.Document;
import java.text.NumberFormat;
class EditWindow extends JFrame implements DocumentListener {
    JTextArea text1,text2;
    Pattern p;
    Matcher m;
    double number=0;
    NumberFormat f=NumberFormat.getInstance();
    JLabel showMessage;
    TreeSet<String> tree=new TreeSet<String>();
    LinkedList<String> mylist=new LinkedList<String>();
    EditWindow(String s) {
        super(s);
        setSize(500,370);
        setLocation(120,120);
        setVisible(true);
        setDefaultCloseOperation(JFrame.EXIT_ON_CLOSE);
        text1=new JTextArea(12,22);
        text2=new JTextArea(12,22);
        text1.setLineWrap(true);
        text1.setWrapStyleWord(true);
        Document document=text1.getDocument();
        【代码1】            //将当前窗口作为文档document发生Document事件时的监视器
        text2.setLineWrap(true);
        text2.setWrapStyleWord(true);
        Box boxV1=Box.createVerticalBox();
        Box boxV2=Box.createVerticalBox();
        boxV1.add(new JLabel("输入一篇英文短文："));
        boxV1.add(new JScrollPane(text1));
        showMessage=new JLabel("短文中使用的单词及频率：");
        boxV2.add(showMessage);
        boxV2.add(new JScrollPane(text2));
        Box boxH=Box.createHorizontalBox();
        boxH.add(boxV1);
        boxH.add(boxV2);
        Container con=getContentPane();
        con.add(boxH,BorderLayout.CENTER);
        con.validate();
        validate();
        f.setMaximumFractionDigits(4);
    }
    【代码2】{                    //定义changeUpdate(DocumentEvent e)方法
        handle();
    }
    【代码3】{                    //定义removeUpdate(DocumentEvent e)方法
        handle();
```

```
        }
     【代码4】{                  //定义insertUpdate(DocumentEvent e)方法
         handle();
     }
     public void handle(){
         text2.setText(null);
         tree.clear();
         mylist.clear();
         String s=text1.getText();
         String pattern="[a-z[A-Z]]+";
         p=Pattern.compile(pattern);
         m=p.matcher(s);
         while(m.find()){
             String str=m.group();
             tree.add(str);
             mylist.add(str);
         }
         Iterator<String> iter=tree.iterator();
         number=tree.size();
         showMessage.setText("短文已经使用了"+(int)number+"个单词");
         int count[]=new int[tree.size()];
         int i=0;
         while(iter.hasNext()){
             count[i]=0;
             String item=iter.next();
             for(int k=0;k<mylist.size();k++) {
                 String temp=mylist.get(k);
                 if(item.equals(temp)) {
                     count[i]++;
                 }
             }
             text2.append("\n"+item+"使用了"+count[i]+"次,使用频率: "+f.format(count[i]/number));
             i++;
         }
     }
 }
 public class SortExample{
     public static void main(String args[]) {
         EditWindow win=new EditWindow("窗口");
     }
 }
```

5. 实验指导与检查

⊙ 文本区可以触发 DucumentEvent 事件，DucumentEven 类在 javax.swing.event 包中。用户在文本区组件的 UI 代表的视图中进行文本编辑操作，使得文本区中的文本内容发生变化，将导致该组件所维护的文档模型中的数据发生变化，从而导致 DucumentEvent 事件的发生。需要使用 addDucumentListener()方法向组件维护的文档注册监视器。监视器需实现 DucumentListener 接口。该接口中有如下 3 个方法：

  public void changedUpdate(DocumentEvent e)

  public void removeUpdate(DocumentEvent e)

  public void insertUpdate(DocumentEvent e)

⊙ 向实验指导教师演示程序的运行效果。

6. 实验报告

实验报告的格式如下（可要求学生填写并由实验指导教师签字）：

学号：_____ 班级：_____ 姓名：_____ 时间：_____

实 验 内 容	回　　答	教师评语
上述程序中的 text2 按字典序显示了 text1 中出现的英文单词。请修改程序，要求 text2 把 text1 中出现的单词，按它们在 text1 中出现的次数（频率）排序		

10.2.5　华容道游戏

1. 实验目的

本实验的目的是学习焦点事件、鼠标事件和键盘事件。

2. 实验要求

华容道是一个传统智力游戏。首先，编写一个按钮的子类，该子类创建的对象代表华容道中的人物。通过焦点事件控制人物的颜色，当人物获得焦点时颜色为蓝色，当失去焦点时颜色为灰色。我们通过键盘事件和鼠标事件来实现曹操、关羽等人物的移动，当人物上发生鼠标事件或键盘事件时，如果鼠标指针的位置是在人物的下方（也就是组件的下半部分）或按键盘上的"↓"键，该人物向下移动。向左、向右和向上的移动原理类似。

图 10-5　华容道

3. 运行效果示例

运行效果如图 10-5 所示。

4. 程序模板

按模板要求，将【代码】替换为程序代码。

MoveExample.java

```
import java.awt.*;
import java.applet.*;
import java.awt.event.*;
import javax.swing.*;
public class MoveExample{
   public static void main(String args[]){
      new Hua_Rong_Road();
   }
}
```

```
}
class Person extends JButton implements FocusListener{
    int number;
    Color c;
    Person(int number,String s){
        super(s);
        this.number=number;
        c=getBackground();
        setFont(new Font("宋体",Font.CENTER_BASELINE,14));
        addFocusListener(this);
    }
    public void focusGained(FocusEvent e){
        setBackground(Color.cyan);
    }
    public void focusLost(FocusEvent e){
        setBackground(c);
    }
}
class Hua_Rong_Road extends JFrame implements KeyListener,MouseListener,ActionListener{
    Person person[]=new Person[10];
    JButton left,right,above,below;
    JButton restart=new JButton("重新开始");
    Container con;
    public Hua_Rong_Road(){
        init();
        setBounds(100,100,320,360);
        setVisible(true);
        validate();
        setDefaultCloseOperation(JFrame.EXIT_ON_CLOSE);
    }
    public void init(){
        con=getContentPane();
        con.setLayout(null);
        con.add(restart);
        restart.setBounds(100,5,120,25);
        restart.addActionListener(this);
        String name[]={"曹操","关羽","张","刘","马","许","兵","兵","兵","兵"};
        for(int i=0;i<name.length;i++) {
            person[i]=new Person(i,name[i]);
            【代码1】            //将当前窗口注册为person[i]的KeyEvent事件监视器
            【代码2】            //将当前窗口注册为person[i]的MouseEvent事件监视器
            con.add(person[i]);
        }
        person[0].setBounds(104,54,100,100);
        person[1].setBounds(104,154,100,50);
```

```
            person[2].setBounds(54, 154,50,100);
            person[3].setBounds(204,154,50,100);
            person[4].setBounds(54, 54, 50,100);
            person[5].setBounds(204, 54, 50,100);
            person[6].setBounds(54,254,50,50);
            person[7].setBounds(204,254,50,50);
            person[8].setBounds(104,204,50,50);
            person[9].setBounds(154,204,50,50);
            person[9].requestFocus();
            left=new JButton();
            right=new JButton();
            above=new JButton();
            below=new JButton();
            con.add(left);
            con.add(right);
            con.add(above);
            con.add(below);
            left.setBounds(49,49,5,260);
            right.setBounds(254,49,5,260);
            above.setBounds(49,49,210,5);
            below.setBounds(49,304,210,5);
            con.validate();
       }
       public void keyPressed(KeyEvent e){
            Person man=【代码3】                              //返回事件源
            if(e.getKeyCode()==KeyEvent.VK_DOWN){
                goDown(man);
            }
            if(e.getKeyCode()==KeyEvent.VK_UP){
                goUp(man);
            }
            if(e.getKeyCode()==KeyEvent.VK_LEFT){
                 goLeft(man);
            }
            if(e.getKeyCode()==KeyEvent.VK_RIGH){
                goRight(man);
            }
       }
       public void keyTyped(KeyEvent e){ }
       public void keyReleased(KeyEvent e){ }
       public void mousePressed(MouseEvent e){
            Person man=【代码4】                              //返回事件源
            int x=-1,y=-1;
            x=e.getX();
            y=e.getY();
            int w=man.getBounds().width;
```

```
            int h=man.getBounds().height;
            if(y>h/2){
                goDown(man);
            }
            if(y<h/2){
                goUp(man);
            }
            if(x<w/2){
                goLeft(man);
            }
            if(x>w/2){
                goRight(man);
            }
        }
        public void mouseReleased(MouseEvent e) { }
        public void mouseEntered(MouseEvent e) { }
        public void mouseExited(MouseEvent e) { }
        public void mouseClicked(MouseEvent e) { }
        public void goDown(Person man){
            boolean move=true;
            Rectangle manRect=man.getBounds();
            int x=man.getBounds().x;
            int y=man.getBounds().y;
            y=y+50;
            manRect.setLocation(x,y);
            Rectangle belowRect=below.getBounds();
            for(int i=0;i<10;i++){
                Rectangle personRect=person[i].getBounds();
                if((manRect.intersects(personRect))&&(man.number!=i)){
                    move=false;
                }
            }
            if(manRect.intersects(belowRect)){
                move=false;
            }
            if(move==true){
                man.setLocation(x,y);
            }
        }
        public void goUp(Person man){
            boolean move=true;
            Rectangle manRect=man.getBounds();
            int x=man.getBounds().x;
            int y=man.getBounds().y;
            y=y-50;
            manRect.setLocation(x,y);
```

```
                Rectangle aboveRect=above.getBounds();
                for(int i=0;i<10;i++){
                    Rectangle personRect=person[i].getBounds();
                    if((manRect.intersects(personRect))&&(man.number!=i)){
                        move=false;
                    }
                }
                if(manRect.intersects(aboveRect)){
                    move=false;
                }
                if(move==true){
                    man.setLocation(x,y);
                }
            }
            public void goLeft(Person man){
                boolean move=true;
                Rectangle manRect=man.getBounds();
                int x=man.getBounds().x;
                int y=man.getBounds().y;
                x=x-50;
                manRect.setLocation(x,y);
                Rectangle leftRect=left.getBounds();
                for(int i=0;i<10;i++){
                    Rectangle personRect=person[i].getBounds();
                    if((manRect.intersects(personRect))&&(man.number!=i)){
                        move=false;
                    }
                }
                if(manRect.intersects(leftRect)){
                    move=false;
                }
                if(move==true){
                    man.setLocation(x,y);
                }
            }
            public void goRight(Person man){
                boolean move=true;
                Rectangle manRect=man.getBounds();
                int x=man.getBounds().x;
                int y=man.getBounds().y;
                x=x+50;
                manRect.setLocation(x,y);
                Rectangle rightRect=right.getBounds();
                for(int i=0;i<10;i++){
                    Rectangle personRect=person[i].getBounds();
                    if((manRect.intersects(personRect))&&(man.number!=i)){
```

```
                move=false;
            }
        }
        if(manRect.intersects(rightRect)){
            move=false;
        }
        if(move==true){
            man.setLocation(x,y);
        }
    }
    public void actionPerformed(ActionEvent e){
        con.removeAll();
        init();
        validate();
        repaint();
    }
}
```

5．实验指导与检查

- 在处理鼠标事件时，程序经常关心鼠标在当前组件坐标系中的位置。鼠标事件调用 getX() 方法返回触发当前鼠标事件时，鼠标指针在事件源坐标系中的 x 坐标；鼠标事件调用 getY() 方法返回触发当前鼠标事件时，鼠标指针在事件源坐标系中的 y 坐标。
- 组件可以触发焦点事件。组件可以使用 **addFocusListener(FocusListener listener)** 方法增加焦点事件监视器。当组件具有焦点监视器后，如果组件从无输入焦点变成有输入焦点，或从有输入焦点变成无输入焦点，都会触发 FocusEvent 事件。创建监视器的类必须实现 FocusListener 接口。该接口有如下 2 个方法：

    public void focusGained(FocusEvent e)

    public void focusLost(FocusEvent e)

    当组件从无输入焦点变成有输入焦点触发 FocusEvent 事件时，监视器调用类实现的接口方法 focusGained(FocusEvent e)；当组件从有输入焦点变成无输入焦点触发 FocusEvent 事件时，监视器调用类实现的接口方法 focusLost(FocusEvent e)。
- 当一个组件处于激活状态时，组件可以成为触发 KeyEvent 事件的事件源。当某个组件处于激活状态时，用户敲击键盘上一个键就导致这个组件触发 KeyEvent 事件。
- 向指导教师演示程序的运行效果。

6．实验报告

实验报告的格式如下（可要求学生填写并由实验指导教师签字）：

学号：_____ 班级：_____ 姓名：_____ 时间：_____

实验内容	回答	教师评语
上述程序中用按钮的名字代表华容道中的人物。按钮上还可以放置图像。请改进程序，使得代表华容道中人物的按钮上都有一幅相关图像		

10.2.6 字体对话框

1. 实验目的

本实验的目的是学习使用对话框的方法。

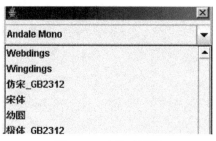

图 10-6 字体对话框

2. 实验要求

单击"设置字体"按钮，弹出一个对话框，然后根据用户选择的字体来改变文本区中的字体。

3. 运行效果示例

运行效果如图 10-6 所示。

4. 程序模板

按模板要求，将【代码】替换为程序代码。

DialogExample.java

```
import java.awt.event.*;
import javax.swing.*;
import java.awt.*;
class FontDialog extends JDialog implements ItemListener,ActionListener{
    JComboBox list;
    JTextArea text;
    Font font;
    JButton yes,cancel;
    JComponent com;
    FontDialog(JComponent com){
        this.com=com;
        【代码1】                                    //将对话框设置为有模式
        yes=new JButton("Yes");
        cancel=new JButton("cancel");
        yes.addActionListener(this);
        cancel.addActionListener(this);
        text=new JTextArea(2,25);
        list=new JComboBox();
        GraphicsEnvironment ge=GraphicsEnvironment.getLocalGraphicsEnvironment();
        String fontName[]=ge.getAvailableFontFamilyNames();
        for(int i=0;i<fontName.length;i++){
            list.addItem(fontName[i]);
        }
        list.addItemListener(this);
        Container con=getContentPane();
        con.setLayout(new FlowLayout());
        Box boxH1=Box.createHorizontalBox(),
            boxH2=Box.createHorizontalBox(),
            boxH3=Box.createHorizontalBox();
```

```java
            Box boxBase=Box.createVerticalBox();
            boxH1.add(list);
            boxH2.add(text);
            boxH3.add(yes);
            boxH3.add(cancel);
            boxBase.add(boxH1);
            boxBase.add(boxH2);
            boxBase.add(boxH3);
            con.add(boxBase);
            setBounds(100,100,280,170);
            setDefaultCloseOperation(JFrame.DISPOSE_ON_CLOSE);
            validate();
        }
        public void itemStateChanged(ItemEvent e){
            String name=(String)list.getSelectedItem();
            font=new Font(name,Font.PLAIN,14);
            text.setFont(font);
            text.setText("hello,奥运");
        }
        public void actionPerformed(ActionEvent e){
            if(e.getSource()==yes){
                com.setFont(font);
                setVisible(false);
            }
            else if(e.getSource()==cancel){
                setVisible(false);
            }
        }
    }
    class Dwindow extends JFrame implements ActionListener{
        JTextArea text;
        JToolBar bar;
        Container con;
        JButton buttonFont;
        Dwindow(){
            buttonFont=new JButton("设置字体");
            text=new JTextArea("显示内容");
            buttonFont.addActionListener(this);
            bar=new JToolBar();
            bar.add(buttonFont);
            con=getContentPane();
            con.add(bar,BorderLayout.NORTH);
            con.add(new JScrollPane(text));
            setBounds(60,60,300,300);
            setVisible(true);
```

```
            validate();
            setDefaultCloseOperation(JFrame.EXIT_ON_CLOSE);
        }
        public void actionPerformed(ActionEvent e){
            if(e.getSource()==buttonFont){
                FontDialog dialog=【代码2】    //创建对话框,并将 text 传递给构造方法的参数
                【代码3】                      //将对话框的 dialog 设置为可见
            }
        }
    }
    public class DialogExample{
        public static void main(String args[]){
            new Dwindow();
        }
    }
```

5. 实验指导与检查

- 通过建立 JDialog 的子类来建立一个对话框类,这个类的一个实例即这个子类创建的一个对象,就是一个对话框。
- 对话框分为无模式和有模式两种。如果一个对话框是有模式的对话框,那么当这个对话框处于激活状态时,只让程序响应对话框内部的事件,程序不能再激活它所依赖的窗口或组件,而且将堵塞当前线程的执行,直到该对话框消失。
- 向实验指导教师演示程序的运行效果。

6. 实验报告

实验报告的格式如下(可要求学生填写并由实验指导教师签字):

学号：_____ 班级：_____ 姓名：_____ 时间：_____

实验内容	回答	教师评语
给上述程序中的对话框增加设置字体、字号(字体大小)的功能		

10.3 知识扩展——计时器

javax.swing 包中提供了一个计时器线程,该计时器的设计原理是基于事件模式,更便于使用。当某些操作需要周期性地执行,就可以使用这个计时器。可以使用 Timer 类的构造方法 Timer(int a, Object b)创建一个计时器。其中,参数 a 的单位是毫秒,确定计时器每隔 a 毫秒"震铃"一次,参数 b 是计时器的监视器。计时器发生的震铃事件是 ActinEvent 类型事件。当震铃事件发生时,监视器就会监视到这个事件,就会执行接口 ActionListener 中的 actionPerformed()方法。因此,当震铃每隔 a 毫秒发生一次时,actionPerformed()方法就被执行一次。若让计时器只震铃一次,可以让计时器调用 setReapeats(boolean b)方法,参数 b 的值取 false 即可。使用 Timer(int a, Object b)方法创建计时器时,对象 b 就自动成为计时器的监视器,不必像其他组件那样(如按钮)使用特定的方法获得监视器,但负责创建监视器的类必须实现接口 Actionlistener。

计时器创建后，使用 Timer 类的 start()方法启动计时器，使用 Timer 类的 stop()方法停止计时器。下面的 Example.java 使用了 Timer 类，该程序中的 3 个按钮不断地随机改变位置。

Example.java
```java
import javax.swing.*;
import java.awt.*;
import java.awt.event.*;
class WindowBox extends JFrame implements ActionListener {
    JButton block1,block2,block3,startOrStop;
    Timer time1,time2,time3;
    int t1,t2,t3;
    int n=0;
    WindowBox() {
        block1=new JButton();
        block2=new JButton();
        block3=new JButton();
        startOrStop=new JButton("start");
        block1.setBackground(Color.red);
        block2.setBackground(Color.blue);
        block3.setBackground(Color.yellow);
        t1=200;
        t2=222;
        t3=160;
        time1=new Timer(t1,this);
        time2=new Timer(t2,this);
        time3=new Timer(t3,this);
        startOrStop.addActionListener(this);
        Container con=getContentPane();
        con.setLayout(null);
        con.add(block1);
        con.add(block2);
        con.add(block3);
        con.add(startOrStop);
        startOrStop.setBounds(100,0,120,30);
        block1.setBounds(0,0,30,30);
        block2.setBounds(260,40,30,30);
        block3.setBounds(23,100,30,30);
        con.validate();
        setBounds(10,12,300,300);
        setVisible(true);
        setDefaultCloseOperation(JFrame.EXIT_ON_CLOSE);
    }
    public void actionPerformed(ActionEvent e) {
        int width=getBounds().width;
        int height=getBounds().height;
```

```java
            if(e.getSource()==time1) {
                int x=(int)(Math.random()*width);
                int y=(int)(Math.random()*height);
                block1.setLocation(x,y);
            }
            if(e.getSource()==time2) {
                int x=(int)(Math.random()*width);
                int y=(int)(Math.random()*height);
                block2.setLocation(x,y);
            }
            if(e.getSource()==time3) {
                int x=(int)(Math.random()*width);
                int y=(int)(Math.random()*height);
                block3.setLocation(x,y);
            }
            if(e.getSource()==startOrStop) {
                n=(n+1)%2;
                if(n==1) {
                    time1.start();
                    time2.start();
                    time3.start();
                    startOrStop.setText("stop");
                }
                else {
                    time1.stop();
                    time2.stop();
                    time3.stop();
                    startOrStop.setText("start");
                }
            }
        }
    }
    public class Example {
        public static void main(String args[]) {
            new WindowBox();
        }
    }
```

第 11 章

Java 中的网络编程

本章导读

- ✿ 知识概述
- ✿ 实验 1　读取服务器中的文件
- ✿ 实验 2　过滤网页中的内容
- ✿ 实验 3　使用套接字传输数据
- ✿ 实验 4　基于 UDP 的图像传输
- ✿ 知识扩展——网络中的数据压缩和传输

11.1 知识概述

URL 类在 java.net 包中。使用 URL（Uniform Resource Locator）创建对象的应用程序称为客户机程序，一个 URL 对象存放着一个具体资源的引用，表明客户机要访问这个 URL 中的资源，利用该 URL 对象可以获取 URL 中的资源。URL 对象调用 InputStream openStream()方法可以返回一个输入流，该输入流指向 URL 对象所包含的资源。通过该输入流，用户可以将服务器上的资源信息读入到客户机。javax.swing 包中的 JEditorPane 类可以解释执行 HTML 文件。也就是说，如果把 URL 中的 HTML 文件读入到 JEditorPane，该 HTML 文件就会被解释执行，显示在 JEditorPane 中，这样就看到了网页的运行效果。

网络套接字是基于 TCP 的有连接通信，套接字连接就是客户机的套接字对象和服务器的套接字对象通过 I/O 流连接在一起。服务器建立 ServerSocket 对象，ServerSocket 对象负责等待客户机请求建立套接字连接，而客户机建立 Socket 对象向服务器发出套接字连接请求。

与基于 TCP 的通信不同，基于 UDP 的通信的信息传递更快，但不提供可靠性保证。也就是说，数据在传输时，用户无法知道数据能否正确到达目的主机，也不能确定数据到达目的地的顺序是否与发送的顺序相同。基于 UDP 通信的基本模式是发送和接收数据包。

11.2 实验练习

11.2.1 读取服务器中的文件

1．实验目的

本实验的目的是结合 I/O 流使用 URL。

图 11-1 读取文件

2．实验要求

创建一个 URL 对象，然后让 URL 对象返回输入流，通过该输入流读取 URL 所包含的资源文件。

3．运行效果示例

运行效果如图 11-1 所示。

4．程序模板

按模板要求，将【代码】替换为程序代码。

ReadFile.java

```
import java.awt.*;
import java.awt.event.*;
import java.net.*;
import java.io.*;
import javax.swing.*;
public class ReadURLSource{
    public static void main(String args[]) {
        new NetWin();
    }
```

```java
}
class NetWin extends JFrame implements ActionListener,Runnable {
    JButton button;
    URL url;
    JTextField inputURLText;                    //输入 URL
    JTextArea area;
    byte b[]=new byte[118];
    Thread thread;
    NetWin(){
        inputURLText=new JTextField(20);
        area=new JTextArea(12,12);
        button=new JButton("确定");
        button.addActionListener(this);
        thread=new Thread(this);
        JPanel p=new JPanel();
        p.add(new JLabel("输入网址："));
        p.add(inputURLText);
        p.add(button);
        add(area,BorderLayout.CENTER);
        add(p,BorderLayout.NORTH);
        setBounds(60,60,560,300);
        setVisible(true);
        validate();
        setDefaultCloseOperation(JFrame.EXIT_ON_CLOSE);
    }
    public void actionPerformed(ActionEvent e){
        if(!(thread.isAlive())){
            thread=new Thread(this);
        }
        try{
            thread.start();
        }
        catch(Exception ee){
            inputURLText.setText("我正在读取"+ url);
        }
    }
    public void run(){
        try{
            int n=-1;
            area.setText(null);
            String name=inputURLText.getText().trim();
            【代码1】                              //使用字符串 name 创建 URL 对象
            String hostName = 【代码2】            //url 调用 getHost()
            int urlPortNumber= url.getPort();
            String fileName=url.getFile();
            InputStream in= 【代码3】              //url 调用方法返回一个输入流
```

```
                    area.append("\n 主机: "+ hostName+ "端口: "+ urlPortNumber+ "包含的文件名字: "+ fileName);
                    area.append("\n 文件的内容如下: ");
                    while((n=in.read(b))!=-1){
                        String s=new String(b,0,n);
                        area.append(s);
                    }
                }
                catch(MalformedURLException e1){
                    inputURLText.setText(""+ e1);
                    return;
                }
                catch(IOException e1){
                    inputURLText.setText(""+ e1);
                    return;
                }
            }
        }
```

5．实验指导与检查

- File 对象 file 调用 toURL()方法可以返回一个以 file 为资源的 URL 对象。
- 向实验指导教师演示程序的运行效果。

6．实验报告

实验报告的格式如下（可要求学生填写并由实验指导教师签字）：

学号：_____ 班级：_____ 姓名：_____ 时间：_____

实 验 内 容	回 答	教 师 评 语
public int getDefaultPort()、public String getRef()、public String getProtocol()等方法都是 URL 对象常用的方法，在模板中，让 url 调用这些方法，并输出这些方法返回的值		

11.2.2　过滤网页中的内容

1．实验目的

本实验的目的是让学生掌握使用 JEditorPane 容器显示 URL 对象的方法。

2．实验要求

不显示网页中的图像，即不显示网页中的<IMG .../>标签。要求程序首先将要显示的网页读入到内存，然后将需要的信息写入到一个临时 HTML 文件中，并创建一个以该临时文件为资源的 URL 对象，最后程序使用 JEditorPane 容器显示这个 URL 对象。

3．运行效果示例

运行效果如图 11-2 所示。

4．程序模板

按模板要求，将【代码】替换为程序代码。

图 11-2 过滤网页中的某些内容

Example.java
```
import javax.swing.*;
import java.awt.*;
import java.awt.event.*;
import java.net.*;
import java.io.*;
import java.util.regex.*;
import javax.swing.event.*;
class Win extends JFrame implements ActionListener,Runnable{
   JButton button;
   URL url, newURL;
   JTextField text;
   JEditorPane editPane;
   byte b[]=new byte[118];
   Thread thread;
   Container con=null;
   JPanel p;
   public Win(){
      text=new JTextField(20);
      editPane= new JEditorPane();
      editPane.setEditable(false);
      button=new JButton("确定");
      button.addActionListener(this);
      thread=new Thread(this);
      p=new JPanel();
      p.add(new JLabel("输入网址："));
      p.add(text);
      p.add(button);
      con=getContentPane();
      con.add(new JScrollPane(editPane),BorderLayout.CENTER);
      con.add(p,BorderLayout.NORTH);
      setBounds(60,60,460,380);
      setVisible(true);
      validate();
      setDefaultCloseOperation(JFrame.EXIT_ON_CLOSE);
```

```
    }
    public void actionPerformed(ActionEvent e){
        if(!(thread.isAlive())){
            thread=new Thread(this);
        }
        try{
            thread.start();
        }
        catch(Exception ee){
            text.setText("我正在读取"+ url);
        }
    }
    public void run(){
        try{
            int m=-1;
            editPane.setText(null);
            url=【代码1】
                     //使用构造方法 URL(String s)创建 url,其中参数 s 由 text 中的文本指定
            InputStream in=【代码2】             //url 返回输入流
            File file=new File("temp.html");
            ByteArrayOutputStream write=new ByteArrayOutputStream();
            while((m=in.read(b))!=-1){
                write.write(b,0,m);
            }
            write.close();
            in.close();
            byte content[]=write.toByteArray();
            String str=new String(content);
            Pattern pattern;
            Matcher match;
            pattern=Pattern.compile("<IMG.*>",Pattern.CASE_INSENSITIVE);
            match=pattern.matcher(str);
            str=match.replaceAll("");
            byte cc[]=str.getBytes();
            ByteArrayInputStream inByte=new ByteArrayInputStream(cc);
            FileOutputStream out=new FileOutputStream(file);
            byte dd[]=new byte[1024];
            while((m=inByte.read(dd,0,1024))!=-1){
                out.write(dd,0,m);
            }
            out.close();
            inByte.close();
            newURL=file.toURL();
            con.removeAll();
            editPane=new JEditorPane();
            editPane.setEditable(false);
```

```
                【代码3】                          //editPane 显示 newURL
                editPane.addHyperlinkListener(new HyperlinkListener(){
                    public void hyperlinkUpdate(HyperlinkEvent e){
                        if(e.getEventType()==HyperlinkEvent.EventType.ACTIVATED){
                            try{
                                URL linkURL=e.getURL();
                                【代码4】                  //editPane 显示 linkURL
                            }
                            catch(IOException e1){
                                editPane.setText(""+ e1);
                            }
                        }
                    }
                });
                con.add(p,BorderLayout.NORTH);
                con.add(new JScrollPane(editPane),BorderLayout.CENTER);
                con.validate();
                validate();
            }
            catch(MalformedURLException e1){
                text.setText(""+ e1);
                return;
            }
            catch(IOException e1){
                text.setText(""+ e1);
                return;
            }
        }
    }
    public class Example{
        public static void main(String args[]){
            new Win();
        }
    }
```

5. 实验指导与检查

- File 对象 file 调用 toURL()方法可以返回一个以 file 为资源的 URL 对象。
- 向实验指导教师演示程序的运行效果。

6. 实验报告

实验报告的格式如下（可要求学生填写并由实验指导教师签字）：

学号：_____ 班级：_____ 姓名：_____ 时间：_____

实验内容	回答	教师评语
给上述程序中增加文本条，用户可以通过该界面输入一个准备过滤掉的 HTML 标签，程序将显示过滤后的网页		

11.2.3 使用套接字传输数据

1．实验目的

本实验的目的是让学生掌握套接字和数据流的使用。

2．实验要求

客户机与服务器建立套接字连接后，服务器向客户机发送一个 1～100 之间的随机数，用户将自己的猜测发送给服务器，服务器向用户发送有关信息："猜大了"、"猜小了"或"猜对了"。

3．运行效果示例

运行效果如图 11-3 和图 11-4 所示。

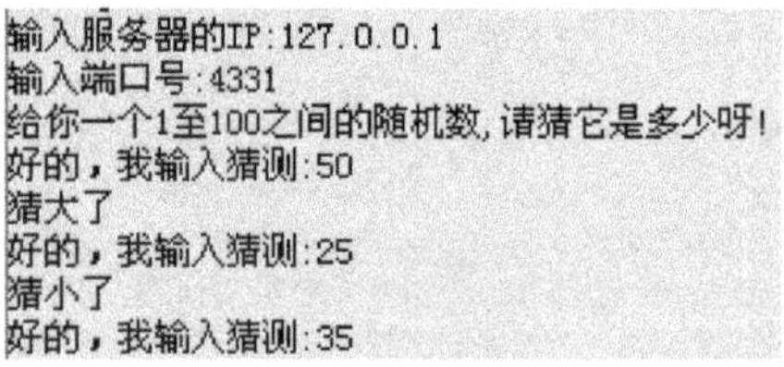

图 11-3　客户机（一） 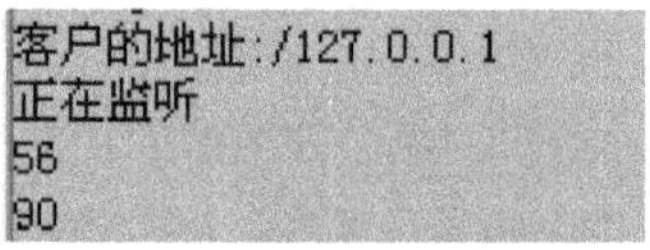　　　　图 11-4　服务器（一）

4．程序模板

按模板要求，将【代码】替换为程序代码。

客户机模板：

ClientGuess.java

```java
import java.io.*;
import java.net.*;
import java.util.*;
public class ClientGuess {
    public static void main(String args[]) {
        Scanner scanner=new Scanner(System.in);
        Socket mysocket=null;
        DataInputStream inData=null;
        DataOutputStream outData=null;
        Thread thread;
        ReadNumber readNumber=null;
        try {
            mysocket=new Socket();
            readNumber=new ReadNumber();
            thread=new Thread(readNumber);              //负责读取信息的线程
            System.out.print("输入服务器的IP: ");
            String IP=scanner.nextLine();
            System.out.print("输入端口号: ");
            int port=scanner.nextInt();
            if(mysocket.isConnected()){ }
            else {
                InetAddress address=InetAddress.getByName(IP);
```

```java
                InetSocketAddress socketAddress=new InetSocketAddress(address,port);
                mysocket.connect(socketAddress);
                InputStream in= 【代码 1】     //mysocket 调用 getInputStream()返回输入流
                OutputStream out= 【代码 2】 //mysocket 调用 getOutputStream()返回输出流
                inData=new DataInputStream(in);
                outData=new DataOutputStream(out);
                readNumber.setDataInputStream(inData);
                readNumber.setDataOutputStream(outData);
                thread.start();                              //启动负责读取随机数的线程
            }
        }
        catch(Exception e) {
            System.out.println("服务器已断开"+ e);
        }
    }
}
class ReadNumber implements Runnable {
    Scanner scanner=new Scanner(System.in);
    DataInputStream in;
    DataOutputStream out;
    public void setDataInputStream(DataInputStream in) {
        this.in=in;
    }
    public void setDataOutputStream(DataOutputStream out) {
        this.out = out;
    }
    public void run(){
        try{
            out.writeUTF("Y");
            while(true) {
                String str=in.readUTF();
                System.out.println(str);
                if(!str.startsWith("询问")){
                    if(str.startsWith("猜对了")) {
                        continue;
                    }
                    System.out.print("好的，我输入猜测： ");
                    int myGuess=scanner.nextInt();
                    String enter=scanner.nextLine();          //消除多余的回车
                    out.writeInt(myGuess);
                }
                else {
                    System.out.print("好的，我输入 Y 或 N: ");
                    String myAnswer=scanner.nextLine();
                    out.writeUTF(myAnswer);
                }
```

                }
            }
            catch(Exception e) {
                System.out.println("与服务器已断开"+ e);
                return;
            }
        }
    }
}
服务器模板：

Server.java
```
    import java.io.*;
    import java.net.*;
    import java.util.*;
    public class ServerNumber {
        public static void main(String args[]) {
            ServerSocket server=null;
            ServerThread thread;
            Socket you=null;
            while(true) {
                try {
                    server= 【代码 3】      //创建在端口 4331 上负责监听的 ServerSocket 对象
                }
                catch(IOException e1) {
                    System.out.println("正在监听");
                }
                try {
                    you= 【代码 4】         //server 调用 accept()返回与客户机相连接的 Socket 对象
                    System.out.println("客户的地址："+ you.getInetAddress());
                }
                catch(IOException e) {
                    System.out.println("正在等待客户");
                }
                if(you!=null) {
                    new ServerThread(you).start();
                }
            }
        }
    }
    class ServerThread extends Thread {
        Socket socket;
        DataInputStream in=null;
        DataOutputStream out=null;
        ServerThread(Socket t) {
            socket=t;
            try {
                out=new DataOutputStream(socket.getOutputStream());
```

```java
                in=new DataInputStream(socket.getInputStream());
            }
            catch (IOException e){ }
        }
        public void run() {
            try {
                while(true) {
                    String str=in.readUTF();
                    boolean boo=str.startsWith("Y")||str.startsWith("y");
                    if(boo) {
                        out.writeUTF("给你一个 1~100 之间的随机数,请猜它是多少? ");
                        Random random=new Random();
                        int realNumber=random.nextInt(100)+1;
                        handleClientGuess(realNumber);
                        out.writeUTF("询问:想继续玩输入 Y,否则输入 N。");
                    }
                    else {
                        return;
                    }
                }
            }
            catch(Exception exp){ }
        }
        public void handleClientGuess(int realNumber) {
            while(true) {
                try {
                    int clientGuess=in.readInt();
                    System.out.println(clientGuess);
                    if(clientGuess>realNumber) {
                        out.writeUTF("猜大了");
                    }
                    else if(clientGuess<realNumber) {
                        out.writeUTF("猜小了");
                    }
                    else if(clientGuess==realNumber) {
                        out.writeUTF("猜对了! ");
                        break;
                    }
                }
                catch(IOException e) {
                    System.out.println("客户离开");
                    return;
                }
            }
        }
    }
}
```

5. 实验指导与检查

- 套接字连接中涉及输入流和输出流操作，客户机或服务器读取数据可能引起堵塞，应该把读取数据放在一个单独的线程中。另外，服务器收到客户机的套接字后，就应该启动一个专门为该客户机服务的线程。
- 向实验指导教师演示程序的运行效果。

6. 实验报告

实验报告的格式如下（可要求学生填写并由实验指导教师签字）：

学号：_____ 班级：_____ 姓名：_____ 时间：_____

实 验 内 容	回　　答	教 师 评 语
改进服务器端程序，能向客户发送用户所猜测的次数		

11.2.4 基于 UDP 的图像传输

1. 实验目的

本实验的目的是让学生掌握 DatagramSocket 类的使用。

2. 实验要求

编写客户机-服务器程序，客户机使用 DatagramSocket 对象将数据包发送到服务器，请求获取服务器的图像。服务器将图像文件包装成数据包并使用 DatagramSocket 对象，将该数据包发送到客户机。先将服务器的程序编译通过并运行，再等待客户机的请求。

3. 运行效果示例

运行效果如图 11-5 和图 11-6 所示。

图 11-5　客户机（二）　　　　图 11-6　服务器（二）

4. 程序模板

按模板要求，将【代码】替换为 Java 程序代码。
客户机模板：

ClientGetImage.java

```
import java.net.*;
import java.awt.*;
import java.awt.event.*;
import java.io.*;
```

```java
import javax.swing.*;
class ImageCanvas extends Canvas{
    Image image=null;
    public ImageCanvas(){
        setSize(200,200);
    }
    public void paint(Graphics g){
        if(image!=null){
            g.drawImage(image,0,0,this);
        }
    }
    public void setImage(Image image){
        this.image=image;
    }
}
public class ClientGetImage extends JFrame implements Runnable,ActionListener{
    JButton b=new JButton("获取图像");
    ImageCanvas canvas;
    ClientGetImage(){
        super("I am a client");
        setSize(320,200);
        setVisible(true);
        b.addActionListener(this);
        add(b,BorderLayout.NORTH);
        canvas=new ImageCanvas();
        add(canvas,BorderLayout.CENTER);
        Thread thread=new Thread(this);
        validate();
        setDefaultCloseOperation(JFrame.EXIT_ON_CLOSE);
        thread.start();
    }
    public void actionPerformed(ActionEvent event){
        byte b[]="请发图像".trim().getBytes();
        try{
            InetAddress address=InetAddress.getByName("127.0.0.1");
            DatagramPacket data= 【代码 1】
                        //创建 data，该数据包的目标地址和端口分别是 address
                        //和 1234，其中的数据为数组 b 的全部字节
            DatagramSocket mailSend= 【代码 2】      //创建负责发送数据的 mailSend 对象
            【代码 3】              //mailSend 发送数据 data
        }
        catch(Exception e){ }
    }
    public void run(){
        DatagramPacket pack=null;
        DatagramSocket mailReceive=null;
```

```
            byte b[]=new byte[8192];
            ByteArrayOutputStream out=new ByteArrayOutputStream();
            try{
                pack=new DatagramPacket(b,b.length);
                mailReceive=【代码4】    //创建在端口5678负责收取数据包的mailReceive对象
            }
            catch(Exception e){ }
            try{
                while(true){
                    mailReceive.receive(pack);
                    String message=new String(pack.getData(),0,pack.getLength());
                    if(message.startsWith("end")){
                        break;
                    }
                    out.write(pack.getData(),0,pack.getLength());
                }
                byte imagebyte[]=out.toByteArray();
                out.close();
                Toolkit tool=getToolkit();
                Image image=tool.createImage(imagebyte);
                canvas.setImage(image);
                canvas.repaint();
                validate();
            }
            catch(IOException e){ }
        }
        public static void main(String args[]){
            new ClientGetImage();
        }
    }
```

服务器模板：

ServerImage.java

```
    import java.net.*;
    import java.io.*;
    public class ServerImage{
        public static void main(String args[]){
            DatagramPacket pack=null;
            DatagramSocket mailReceive=null;
            ServerThread thread;
            byte b[]=new byte[8192];
            InetAddress address=null;
            pack=new DatagramPacket(b,b.length);
            while(true){
                try{
                    mailReceive= new DatagramSocket(1234);
                }
```

```java
            catch(IOException e1){
                System.out.println("正在等待");
            }
            try{
                mailReceive.receive(pack);
                address=pack.getAddress();
                System.out.println("客户的地址: "+ address);
            }
            catch(IOException e){ }
            if(address!=null){
                new ServerThread(address).start();
            }
        }
    }
}
class ServerThread extends Thread{
    InetAddress address;
    DataOutputStream out=null;
    DataInputStream in=null;
    String s=null;
    ServerThread(InetAddress address){
        this.address=address;
    }
    public void run(){
        FileInputStream in;
        byte b[]=new byte[8192];
        try{
            in=new FileInputStream ("a.jpg");
            int n=-1;
            while((n=in.read(b))!=-1){
                DatagramPacket data=new DatagramPacket(b,n,address,5678);
                DatagramSocket mailSend=new DatagramSocket();
                mailSend.send(data);
            }
            in.close();
            byte end[]="end".getBytes();
            DatagramPacket data=new DatagramPacket(end,end.length,address,5678);
            DatagramSocket mailSend=new DatagramSocket();
            mailSend.send(data);
        }
        catch(Exception e){ }
    }
}
```

5. 实验指导与检查

⊙ 与基于 TCP 的数据传输不同，基于 UDP 的数据传输更快，但不提供可靠性保证。也就是说，UDP 数据在传输时，用户无法知道数据能否正确到达目的地主机，也不能确定数据到达目的地的顺序是否与发送的顺序相同。
- 基于 UDP 数据传输的基本模式是：创建数据包，然后将数据包发往目的地；接收数据包，然后查看数据包中的内容。
- 向实验指导教师演示程序的运行效果。

6．实验报告

实验报告的格式如下（可要求学生填写并由实验指导教师签字）：

学号：_____ 班级：_____ 姓名：_____ 时间：_____

实 验 内 容	回　　答	教 师 评 语
将上述程序改成用户从服务器获取一个文本文件的内容，并显示在客户机上		

11.3 知识扩展——网络中的数据压缩和传输

由于 Internet 带宽是有限的，将数据压缩后再传输是一个好办法。

当使用套接字进行网络通信时，可以使用 ZipOutputStream 流进行数据压缩，将套接字返回的输出流作为 ZipOutputStream 流的底层流，然后 ZipOutputStream 流将数据压缩到底层流再发送到目的地。ZipOutputStream 流可以将若干个文件压缩到底层流再发送到目的地。先使用 ZipOutputStream 类的构造方法 public ZipOutputStream(OutputStream out)创建一个对象，该对象以套接字返回的输出流作为目的地，即作为该流的底层流，如

　　　　ZipOutputStream out=new ZipOutputStream(socket.getOutputStream());

再让 ZipOutputStream 对象 out 调用

　　　　public void putNextEntry(ZipEntry e)

方法类确定向底层流写入下一个文件的位置，如

　　　　out.putNextEntry(new ZipEntry("A.txt"));

然后，out 对象调用 write()方法将文件 A.txt 压缩后写入到底层流中。

数据接收方对应地使用 ZipInputStream 类创建对象，该对象以套接字获取的输入流作为源，即作为该流的底层流，如

　　　　ZipInputStream in=new ZipInputStream(socket.getInputStream());

然后让 ZipInputStream 对象 in 找到发送方写入到底层流中的下一个文件，如

　　　　ZipEntry zipEntry=in.getNextEntry();

那么，in 调用 read()方法可以读取（解压）找到的该文件。

下面的 Server.java 通过套接字将两个文件压缩后发送给客户机。

服务器代码如下：

Server.java

```
import java.io.*;
import java.net.*;
import java.util.zip.*;
public class Server{
```

```java
        public static void main(String args[]){
            ServerSocket server=null;
            ServerThread thread;
            Socket you=null;
            while(true){
                try{
                    server=new ServerSocket(4331);
                }
                catch(IOException e1){
                    System.out.println("正在监听");
                }
                try{
                    you=server.accept();
                    System.out.println("客户的地址: "+you.getInetAddress());
                }
                catch(IOException e){
                    System.out.println("正在等待客户");
                }
                if(you!=null){
                    new ServerThread(you).start();
                }
                else{
                    continue;
                }
            }
        }
    }
    class ServerThread extends Thread{
        Socket socket;
        ZipOutputStream out;
        String s=null;
        ServerThread(Socket t){
            socket=t;
            try{
                out=new ZipOutputStream(socket.getOutputStream());
            }
            catch(IOException e){ }
        }
        public void run(){
            try{
                out.putNextEntry(new ZipEntry("Example.java"));
                FileInputStream reader=new FileInputStream("Example.java");
                byte b[]=new byte[1024];
                int n=-1;
                while((n=reader.read(b,0,1024))!=-1){
                    out.write(b,0,n);              //发送压缩后的数据到客户机
```

```
            }
            out.putNextEntry(new ZipEntry("E.java"));
            reader=new FileInputStream("E.java");
            n=-1;
            while((n=reader.read(b,0,1024))!=-1)
               out.write(b,0,n);                    //发送压缩后的数据到客户机
            }
            reader.close();
            out.close();
        }
        catch(IOException e){ }
    }
}
```

下面的 Client.java 通过套接字接收服务器发来的数据。

客户机代码如下：

Client.java

```
    import java.net.*;
    import java.io.*;
    import java.awt.*;
    import java.awt.event.*;
    import javax.swing.*;
    import java.util.zip.*;
    class Client extends JFrame implements Runnable,ActionListener{
        JButton connection,getFile;
        JTextArea showResult;
        Socket socket=null;
        ZipInputStream in;
        Thread thread;
        public Client(){
            socket=new Socket();
            connection=new JButton("连接服务器，获取文件内容");
            Container con=getContentPane();
            con.setLayout(new FlowLayout());
            showResult=new JTextArea(10,28);
            con.add(connection);
            con.add(new JScrollPane(showResult));
            connection.addActionListener(this);
            thread=new Thread(this);
            setBounds(100,100,460,410);
            setVisible(true);
            setDefaultCloseOperation(JFrame.EXIT_ON_CLOSE);
        }
        public void run(){
            byte b[]=new byte[1024];
            ZipEntry zipEntry=null;
```

```java
                while(true){
                    try{
                        while((zipEntry=in.getNextEntry())!=null){
                            showResult.append("\n"+ zipEntry.toString()+ ": \n");
                            int n=-1;
                            while((n=in.read(b,0,1024))!=-1){
                                String str=new String(b,0,n);
                                showResult.append(str);
                            }
                        }
                    }
                    catch(IOException e){ }
                }
            }
            public void actionPerformed(ActionEvent e){
                if(e.getSource()==connection){
                    try{
                        if(socket.isConnected()){ }
                        else{
                            InetAddress address=InetAddress.getByName("127.0.0.1");
                            InetSocketAddress socketAddress=new InetSocketAddress(address,4331);
                            socket.connect(socketAddress);
                            in=new ZipInputStream(socket.getInputStream());
                            thread.start();
                        }
                    }
                    catch(IOException ee){
                        System.out.println(ee);
                    }
                }
            }
            public static void main(String args[]){
                Client win=new Client();
            }
        }
```

第 12 章

Java 数据库操作

本章导读

- ✿ 知识概述
- ✿ 实验 1　JDBC-ODBC 桥接器
- ✿ 实验 2　查询、更新和插入操作
- ✿ 实验 3　预处理语句
- ✿ 实验 4　事务处理
- ✿ 知识扩展——MySQL 简介

12.1 知识概述

JDBC（Java DataBase Connectivity）是 Java 运行平台的核心类库中的一部分，提供了访问数据库的 API，它由一些 Java 类和接口组成。在 Java 中，JDBC 可以实现对数据库中表记录的查询、修改和删除等操作。JDBC 技术在数据库开发中占有重要的地位，JDBC 操作不同的数据库仅仅是连接方式上的差异而已，使用 JDBC 的应用程序一旦与数据库建立连接，就可以使用 JDBC 提供的 API 操作数据库。

1. JDBC-ODBC 桥接器

使用 JDBC-ODBC 桥接器方式的机制是，应用程序只需建立 JDBC 与 ODBC 之间的连接，即建立 JDBC-ODBC 桥接器，而与数据库的连接由 ODBC 去完成。ODBC 使用"数据源"来管理数据库，所以必须事先将某个数据库设置成 ODBC 所管理的一个数据源，应用程序只能请求与 ODBC 管理的数据源建立连接。

2. 查询、更新与插入操作

JDBC 与数据库表进行交互的主要方式是使用 SQL 语句。JDBC 提供的 API 可以将标准的 SQL 语句发送给数据库，实现与数据库的交互。

3. 预处理

预处理就是 Java 应用程序通过 PreparedStatement 对象，将 SQL 语句解释为数据库底层的内部命令，然后直接让数据库去执行这个命令，不仅可以减轻数据库的负担，而且提高了 Java 应用程序访问数据库的速度。

4. 事务处理

事务是保证数据库中数据完整性和一致性的重要机制。事务由一组 SQL 语句组成。事务处理是指：应用程序保证事务中的 SQL 语句要么全部都执行，要么一个都不执行。

12.2 实验练习

12.2.1 JDBC-ODBC 桥接器

1. 实验目的

本实验的目的是让学生学会如何使用 JDBC-ODBC 桥接器方式去访问 Excel 电子表格。

2. 实验要求

按照建立 JDBC-ODBC 桥接器方式与 Excel 电子表格建立连接，并查询有关数据。

3. 运行效果示例

运行效果如图 12-1 所示。

```
C:\1000>java Excel
a123,电视,台,2900.0
b123,冰箱,台,3400.0
c324,电视,台,3600.0
```

图 12-1 查询电子表格

4. 程序模板与操作步骤

本实验主要训练建立 JDBC-ODBC 桥接器的步骤，请按要求的步骤进行操作。

（1）创建电子表格

打开 Microsoft Excel，建立如图 12-2 所示的电子表格。

	A	B	C	D	E	F	G
1	货号	品名	单位	单价(元)	库存量	生产者	质量担保期限
2	a123	电视	台	2900	4	长城公司	3年
3	b123	冰箱	台	3400	5	里泊公司	5年
4	c324	电视	台	3600	3	佳美集团	3年
5	D23	PC电脑	台	9088	12	IBM	2年

图 12-2　电子表格 goods.xls

（2）设置表

与访问数据库不同的是，必须在电子表格中选取一工作区作为连接时使用的表。在 Excel 电子表格中，拖动鼠标选出范围，然后在 Excel 菜单中选择"插入"→"名称"→"定义"命令，给选中的工作区命名为 message。现在就有了一个名字是 message、有 4 个字段（货号、品名、单位和单价（元））的表，包含 3 条记录，如图 12-2 所示。

（3）保存电子表格

将电子表格命名为 goods.xsl，保存到 D:\ch12 目录中。

（4）设置数据源

设置好 message 表并保存电子表格 goods.xls 之后，开始设置数据源。为数据源选择的驱动程序必须是 Microsoft Excel Driver，要求设置的数据源的名字是 star。在设置数据源时，单击其中的"选项"，将"只读"属性设置为未选中状态。

（5）将下列代码编译、运行，效果如图 12-1 所示。

Excel.java

```java
import java.sql.*;
public class Excel{
    public static void main(String args[]){
        Connection con;
        Statement sql;
        ResultSet rs;
        try{
            Class.forName("sun.jdbc.odbc.JdbcOdbcDriver");
        }
        catch(ClassNotFoundException e){
            System.out.println(""+ e);
        }
        try{
            con=DriverManager.getConnection("jdbc:odbc:star","","");
            sql=con.createStatement();
            rs=sql.executeQuery("SELECT * FROM message");
            while(rs.next()){
                String number=rs.getString(1);
                String name=rs.getString(2);
```

```
                String unit=rs.getString(3);
                double price=rs.getInt(4);
                System.out.println(number+ ","+ name+ ","+ unit+ ","+ price);
            }
            con.close();
        }
        catch(SQLException e){
            System.out.println(e);
        }
    }
}
```

5．实验指导与检查

- 选择"控制面板"→"管理工具"→"ODBC 数据源"命令来设置数据源。
- 应用程序所在的计算机负责创建数据源，即将本地或远程计算机上的数据库设置成自己要访问的数据源，因此必须保证应用程序所在计算机有 ODBC 系统。

6．实验报告

实验报告的格式如下（可要求学生填写并由实验指导教师签字）：

学号：_____　班级：_____　姓名：_____　时间：_____

实 验 内 容	回　　答	教 师 评 语
在 Excel.java 中增加更新和插入操作的 SQL 语句		
在电子表格中定义出两个以上的表，并查询所定义的表		

12.2.2　查询、更新和插入操作

1．实验目的

本实验的目的是掌握使用 JDBC 提供的 API 与数据库交互信息的方法，如查询、更新和插入等操作。

2．实验要求

（1）使用 Microsoft Access 创建名字为 car.mdb 的数据库，然后从中创建名为 list 的表。list 表的结构为：number（文本，主键），name（文本），price（数字，双精度），madeDate（日期）。

（2）将 car.mdb 数据库设置为 ODBC 数据源，该数据源的名字是 mycar。

（3）编写应用程序，向 list 表插入记录、查询这些记录，然后更新其中的某些记录，并查询更新后的记录。

3．运行效果示例

运行效果如图 12-3 所示。

```
C:\1000>java LookCar
001,奔驰,390000.0,2008-12-20
002,奥迪,320000.0,2009-07-20
003,宝马,350000.0,2009-10-10
001,奔驰,380000.0,2008-12-20
```

图 12-3　操作表

4．程序模板

按模板要求，将【代码】替换为程序代码。

LookCar.java
```
import java.sql.*;
public class LookCar{
    public static void main(String args[]){
        Connection con;
        Statement sql;
        ResultSet rs;
        try{
            Class.forName("sun.jdbc.odbc.JdbcOdbcDriver");
        }
        catch(ClassNotFoundException e){
            System.out.println(""+ e);
        }
        try{
            con=DriverManager.getConnection("jdbc:odbc:mycar","","");
            sql=con.createStatement();
            【代码1】            //向 list 表插入记录：('001','奔驰',390000,'2008-12-20')
            【代码2】            //向 list 表插入记录：('002','奥迪',320000,'2009-07-20')
            【代码3】            //向 list 表插入记录：('003','宝马',350000,'2009-10-10')
            rs=sql.executeQuery("SELECT * FROM list");
            while(rs.next()){
                String number=rs.getString(1);
                String name=rs.getString(2);
                double price=rs.getDouble(3);
                Date madeDate=rs.getDate(4);
                System.out.println(number+ ","+ name+ ","+ price+ ","+ madeDate);
            }
            【代码4】            //将 list 表中 number 是 001 的 price 的值更新为 380000
            rs=【代码5】         //查询 number 为 001 的记录
            boolean boo=rs.next();
            if(boo){
                String number=rs.getString(1);
                String name=rs.getString(2);
                double price=rs.getDouble(3);
                Date madeDate=rs.getDate(4);
                System.out.println(number+ ","+ name+ ","+ price+ ","+ madeDate);
            }
            con.close();
        }
        catch(SQLException e){
            System.out.println(e);
        }
    }
}
```

5．实验指导与检查

⊙ 设置数据源时，如果提示"非法路径"，则关闭 Micsosoft Access 打开的 student.mdb 数据库。

- 结果集 ResultSet 对象调用 next()方法可以顺序查询表中的记录。结果集对象将游标最初定位在第一行的前面，第一次调用 next()方法使游标移动到第一行。next()方法返回一个 boolean 类型数据，游标移动到最后一行后返回 false。
- 向实验指导教师演示程序的运行效果。

6．实验报告

实验报告的格式如下（可要求学生填写并由实验指导教师签字）：

学号：_____　　班级：_____　　姓名：_____　　时间：_____

实 验 内 容	回　　答	教 师 评 语
查询价格在一定范围的记录		
将 number 是 002 的 madeDate 的值更新		
按价格排序记录		
按生产日期排序记录		
按名字模糊查询		

12.2.3　预处理语句

1．实验目的

本实验的目的是让学生学会使用预处理语句的方法。

2．实验要求

（1）所用数据库与 12.2.2 节中的数据库相同。
（2）使用预处理语句查询记录、插入记录和更新记录。

3．运行效果示例

运行效果如图 12-4 所示。

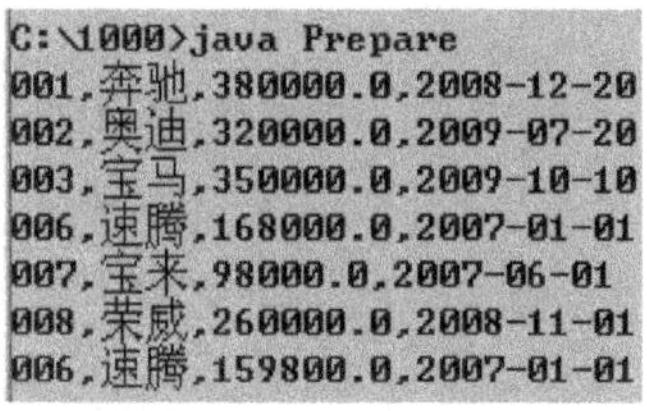

4．程序模板

按模板要求，将【代码】替换为程序代码。

图 12-4　使用预处理语句

Prepare.java

```java
import java.sql.*;
import java.util.Calendar;
public class Prepare {
    public static void main(String args[]){
        Connection con;
        PreparedStatement pre;
        ResultSet rs;
        try{
            Class.forName("sun.jdbc.odbc.JdbcOdbcDriver");
        }
        catch(ClassNotFoundException e){
            System.out.println(""+ e);
        }
```

```
Calendar calendar=Calendar.getInstance();
calendar.set(2007,0,1);                    //注意，0代表一月
Date date1=new Date(calendar.getTimeInMillis());
calendar.set(2007,5,1);
Date date2=new Date(calendar.getTimeInMillis());
calendar.set(2008,10,1);
Date date3=new Date(calendar.getTimeInMillis());
String[] number={"006","007","008"};
String[] name={"速腾","宝来","荣威"};
double[] price={168000,98000,260000};
Date [] date={date1,date2,date3};
try{
   con=DriverManager.getConnection("jdbc:odbc:mycar","","");
   String sql="SELECT * FROM list";
   pre=【代码1】        //con 调用 prepareStatement()方法，sql 作为参数返回一个预处理语句
   rs=pre.executeQuery();
   while(rs.next()){
      String nu=rs.getString(1);
      String na=rs.getString(2);
      double pr=rs.getDouble(3);
      Date madeDate=rs.getDate(4);
      System.out.println(nu+ ","+ na+ ","+ pr+ ","+ madeDate);
   }
   pre=con.prepareStatement("INSERT INTO list VALUES(?,?,?,?)");
   for(int i=0;i<number.length;i++){
      【代码2】        //pre 调用方法将条件中第1个 "?" 的值设置为 number[i]
      【代码3】        //pre 调用方法将条件中第2个 "?" 的值设置为 name[i]
      【代码4】        //pre 调用方法将条件中第3个 "?" 的值设置为 price[i]
      【代码5】        //pre 调用方法将条件中第4个 "?" 的值设置为 date[i]
      pre.executeUpdate();
   }
   pre=con.prepareStatement("SELECT *  FROM list  WHERE number>=?");
   【代码6】        //pre 调用方法将条件中 "?" 的值设置为 number[0]
   rs=pre.executeQuery();
   while(rs.next()) {
      String nu=rs.getString(1);
      String na=rs.getString(2);
      double pr=rs.getDouble(3);
      Date madeDate=rs.getDate(4);
      System.out.println(nu+ ","+ na+ ","+ pr+ ","+ madeDate);
   }
   pre=con.prepareStatement("UPDATE list  SET price=?  WHERE number=?");
   【代码7】        //pre 调用方法将条件中第1个 "?" 的值设为 159800
   pre.setString(2,number[0]);
                   //pre 调用方法将条件中第2个 "?" 的值设为 number[0]
   pre.executeUpdate();
```

```
            pre=con.prepareStatement("SELECT *  FROM list WHERE number='"+number[0]+"'");
            rs=pre.executeQuery();
            boolean boo=rs.next();
            if(boo){
                String nu=rs.getString(1);
                String na=rs.getString(2);
                double pr=rs.getDouble(3);
                Date madeDate=rs.getDate(4);
                System.out.println(nu+ ","+ na+ ","+ pr+ ","+ madeDate);
            }
            con.close();
        }
        catch(SQLException e){
            System.out.println(e);
        }
    }
}
```

5. 实验指导与检查

- 在对 SQL 进行预处理时,可以使用通配符 "?" 来代替字段的值,只要在预处理语句执行之前再设置通配符所表示的具体值即可。
- PreparedStatement 对象可以重复执行,提高了访问数据库的速度。
- 向实验指导教师演示程序的运行效果。

6. 实验报告

实验报告的格式如下(可要求学生填写并由实验指导教师签字):

学号:_____ 班级:_____ 姓名:_____ 时间:_____

实验内容	回答	教师评语
使用预处理语句删除一条记录		
使用预处理语句按日期排序记录		
使用预处理语句按价格排序记录		

12.2.4 事务处理

1. 实验目的

本实验的目的是掌握事务的构成和事务处理。

2. 实验要求

(1) 使用 Microsoft Access 创建名字是 buybook.mdb 的数据库,然后在数据库中创建名字分别是 person 和 orderform 的表。

person 表的结构为:userId(文本,主键), address(文本), userMoney(数字,双精度)。

orderform 表的结构为:userId(文本,外键), address(文本), bookname(文本), bookprice(数字,双精度)。

用 Access 打开 person 表,录入 "001,'北京万寿路 65 号', 98" 等若干条记录。

（2）将 buybook.mdb 数据库设置为 ODBC 数据源，该数据源的名字是 book。

（3）编写应用程序，将 person 表中某记录 L 的 userMoney 的值减去 n，然后在 orderform 中插入一条记录，该记录的 bookprice 的值是 n，其余字段值与记录 L 对应的字段值相同。

（4）在 Microsoft Access 管理系统的"工具"菜单中选择"关系"命令，将 person 表与 orderform 表关联，设置 user 表的 userId 与 orderform 表中的 userId 是"一对多"关系。

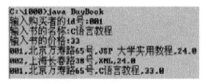

图 12-5 事务处理

3．运行效果示例

运行效果如图 12-5 所示。

4．程序模板

按模板要求，将【代码】替换为 Java 程序代码。

BuyBook.java

```
import java.sql.*;
import java.util.*;
public class BuyBook{
    public static void main(String args[]){
        Connection con=null;
        PreparedStatement pre;
        ResultSet rs;
        try{
            Class.forName("sun.jdbc.odbc.JdbcOdbcDriver");
        }
        catch(ClassNotFoundException e){
            System.out.println(""+ e);
        }
        try{
            Scanner scan=new Scanner(System.in);
            System.out.print("输入购买者的 ID: ");
            String id=scan.nextLine();
            System.out.print("输入书的名称: ");
            String bookname=scan.nextLine();
            System.out.print("输入书的价格: ");
            double n=scan.nextDouble();
            con=DriverManager.getConnection("jdbc:odbc:book","","");
            【代码1】                          //关闭自动提交模式
            pre=con.prepareStatement("SELECT *  FROM person  WHERE userId='"+ id+ "'");
            rs=pre.executeQuery();
            rs.next();
            id=rs.getString("userId");
            String address=rs.getString("address");
            double money=rs.getDouble("userMoney");
            money=money-n;
            if(money<0){
                【代码2】                      //开始事务处理
                con.close();
```

```
            }
            else{
                pre=con.prepareStatement("UPDATE person SET userMoney=? WHERE userId='"+ id+ "'");
                pre.setDouble(1,money);
                pre.executeUpdate();
                pre=con.prepareStatement("INSERT INTO orderform  VALUES(?,?,?,?)");
                pre.setString(1,id);
                pre.setString(2,address);
                pre.setString(3,bookname);
                pre.setDouble(4,n);
                pre.executeUpdate();
                 【代码3】                                 //开始事务处理
                con.close();
            }
            con=DriverManager.getConnection("jdbc:odbc:book","","");
            Statement sql=con.createStatement();
            rs=sql.executeQuery("SELECT * FROM orderform");
            while(rs.next()){
                id=rs.getString(1);
                address=rs.getString(2);
                String bookName=rs.getString(3);
                double bookPrice=rs.getDouble(4);
                System.out.println(id+ ","+ address+ ","+ bookName+ ","+ bookPrice);
            }
            con.close();
        }
        catch(SQLException e){
            try{
                【代码4】                                 //撤销事务所做的操作
            }
            catch(SQLException exp){}
            System.out.println(e);
        }
    }
}
```

5. 实验指导与检查

关闭自动提交模式后，必须进行事务处理。

6. 实验报告

实验报告的格式如下（可要求学生填写并由实验指导教师签字）：

学号：_____ 班级：_____ 姓名：_____ 时间：_____

实验内容	回　　答	教师评语
在数据库book中再增加一个表，该表是库存表，当用户购买一本书之后，将库存表中相应字段值更新		

12.3 知识扩展——MySQL 简介

MySQL 是目前比较流行的一种网络数据库,尽管是开源项目,但功能强大、不依赖于平台,受到广泛的关注。目前,ODBC 不支持 MySQL 数据库,因此不能使用 JDBC-ODBC 桥接器方式与 MySQL 数据库建立连接,只能使用加载 MySQL 的纯 Java 驱动程序来与 MySQL 数据库建立连接。本节先介绍 MySQL 数据库服务器的安装和使用,然后介绍怎样与 MySQL 数据库建立连接。

1. 安装 MySQL 数据库管理系统

MySQL 是开源项目,很多网站提供免费下载。可以使用任何搜索引擎搜索关键字"MySQL 下载"来获得有关的下载地址,推荐 MySQL 的官方网站 http://www.mysql.com,该网站免费提供 MySQL 最新版本的下载和相关技术文章。本书下载的版本是 mysql-5.0.22-win32.zip,该版本可以安装在 Windows 操作系统平台上(如果是其他操作系统平台,请按网站的提示下载相应的版本)。

mysql-5.0.22-win32.zip 的安装步骤如下:

(1) 将 mysql-5.0.22-win32.zip 解压到某个目录,如 E:\mysql。
(2) 双击 E:\mysql 下的 setup.exe 文件开始安装。
(3) 出现安装界面,在该界面上单击"Next"按钮。
(4) 出现选择安装方式界面,有 3 种方式:Typical、Complete 和 Custom,默认是 Typical。这里选择 Custom,因为 Typical 和 Complete 方式不允许改变安装路径,只能安装在 C 盘。
(5) 出现选择 MySQL 功能和安装路径的界面。由于需要 MySQL 是一个网络数据库,即是一个数据库服务器,所以选择"MySQL Server",这也是默认选择。单击该界面上的"Change"按钮,选择安装路径,这里选择的安装路径是"D:\MySQL Server 5.0"。
(6) 出现确认安装界面。开始安装,会出现安装进度条,只要按界面提示进行操作即可。
(7) 出现是否需要在 mysql.com 上注册的界面,这里选择跳过,即单击"Skip Sign-Up",然后单击"Next"按钮。
(8) 出现确认安装结束界面。将该界面上是否配置 MySQL 选择"否",即使用默认配置,单击"确认"按钮,完成安装过程。

2. 启动 MySQL 数据库服务器

MySQL 数据库安装后会有如图 12-6 所示的目录结构。

为了启动 MySQL 数据库服务器,需要执行 MySQL 安装目录下 bin 子目录中的 mysqld.exe 文件。打开 MS-DOS 命令行窗口,并使用 MS-DOS 命令进入到 bin 目录中:

  cd D:\MySQL Server 5.0\bin

然后在命令行输入"mysqld-nt"(如图 12-7 所示)来启动 MySQL 数据库服务器。如果启动成功,MySQL 数据库服务器将占用当前 MS-DOS 窗口。

图 12-6 MySQL 安装目录结构 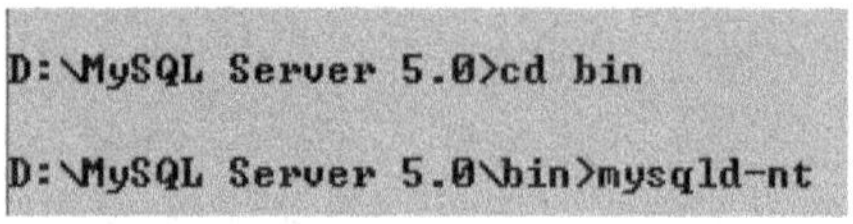   图 12-7 启动 MySQL 服务器

3．启动 MySQL 监视器

启动 MySQL 监视程序可以实现创建数据库、建表等操作。为了启动 MySQL 监视程序，需要执行 MySQL 安装目录下 bin 子目录中的 mysql.exe 文件。再打开 MS-DOS 命令行窗口，并使用 MS-DOS 命令进入到 bin 目录中，然后使用默认的 root 用户启动 MySQL 监视器（在安装 MySQL 时，root 用户是默认的一个用户，没有密码）。命令如下：

```
mysql -h localhost -u root -p
```

如果在安装 MySQL 时设置了密码（本机设置的密码是 123），按要求输入密码即可。

如果没有设置密码，输入命令“mysql –u root”来启动 MySQL 监视器。成功启动 MySQL 监视器后，MS-DOS 窗口出现“mysql>”，如图 12-8 所示。如果想关闭 MySQL 监视器，输入“exit”命令即可。

```
D:\MySQL Server 5.0\bin>mysql -h localhost -u root -p
Enter password: ***
Welcome to the MySQL monitor.  Commands end with ; or \g.
Your MySQL connection id is 6 to server version: 5.0.22-community-nt

Type 'help;' or '\h' for help. Type '\c' to clear the buffer.

mysql> _
```

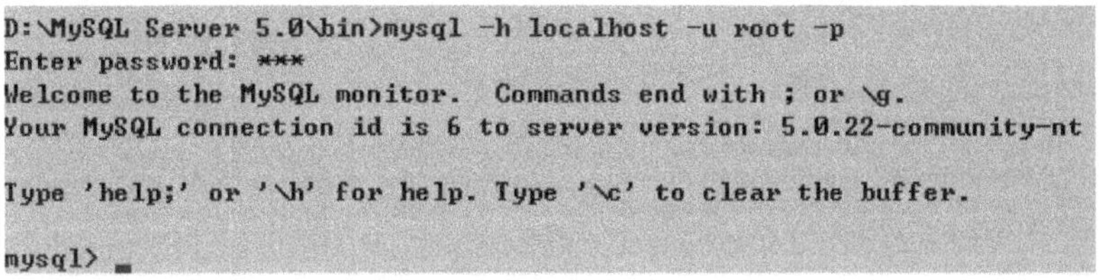

图 12-8　启动 MySQL 监视器

4．创建数据库

启动 MySQL 监视器后，就可以使用 SQL 语句创建数据库、建表等操作。也可以下载相应的图形界面的 MySQL 管理工具进行创建数据库、建表等操作，这些 MySQL 管理工具有免费的，也有需要购买的。本节主要介绍怎样在 Java 中与 MySQL 建立连接，所以采用在 MS-DOS 命令行窗口中输入 SQL 语句建立数据库和创建表。

MySQL 要求 SQL 语句必须用“;”结束，在编辑 SQL 语句的过程中可以使用“\c”终止当前 SQL 语句的编辑。

下面使用 MySQL 监视器创建一个名字为 person 的数据库。创建数据库的命令格式为：

```
create database 数据库名;
```

如图 12-9 所示。

```
mysql> create database person;
Query OK, 1 row affected (0.45 sec)

mysql> _
```

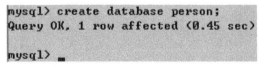

图 12-9　创建数据库

如果创建成功，该数据库将被保存在 MySQL 安装目录的 data 子目录中，数据库的扩展名是 .frm。

5．为数据库建表

创建数据库后，就可以使用 SQL 语句在该库中进行建表操作。为了在某个数据库中建表，必须首先进入该数据库，命令为：

```
use 数据库名
```

进入数据库 person 的操作如图 12-10 所示。进入数据库之后，可以使用如下命令创建表：

```
create table 表名(
```

```
字段  类型（大小）是否允许空值,
字段  类型（大小）是否允许空值,
);
```

下面建立一个名为 message 的表，该表的字段为 number（varchar 类型）、name（varchar 类型）、birthday（date 类型）、height（float 类型）。建表操作如图 12-11 所示。建表之后，就可以使用 SQL 语句对表进行添加、更新和查询操作。现在向表 message 中添加 1 条记录，操作过程如图 12-12 所示。

```
mysql> use person
Database changed
```

```
mysql> create table message(
    -> number varchar(20) not null,
    -> name varchar(50),
    -> birthday date,
    -> height double,
    -> Primary Key(number)
    -> );
Query OK, 0 rows affected (0.13 sec)
```

图 12-10 进入数据库 　　　　　　　图 12-11 在数据库中建表

```
mysql> insert into message values('001','zhangsan','1995-12-26',1.78);
Query OK, 1 row affected (0.00 sec)
```

图 12-12 向表中插入记录

6．使用纯 Java 数据库驱动程序连接 MySQL 数据库

前面介绍了怎样安装 MySQL Server 以及怎样建立数据库和表，现在讲述 Java 应用程序怎样与 MySQL Server 数据库管理系统管理的数据库建立连接。

为了能与 MySQL 数据库服务器管理的数据库建立连接，首先必须保证该 MySQL 数据库服务器已经启动，如果没有更改过 MySQL 数据库服务器的配置，那么该数据库服务器占用的端口是 3306。

Java 应用程序所在计算机必须安装能与 MySQL Server 建立连接的驱动程序，否则无法建立连接。

（1）加载纯 Java 驱动程序

登录 MySQL 官方网站 http://www.mysql.com 下载驱动程序，如 mysql-connector-java-5.0.4.zip，将其解压至硬盘，则 mysql-connector-java-5.0.4-bin.jar 文件就是连接 MySQL 数据库的纯 Java 驱动程序。将它复制到 Tomcat 服务器所使用的 JDK 的\jre\lib\ext 文件夹中，如 C:\jdk1.6\jre\lib\ext。

应用程序加载 SQL Server 驱动程序的代码如下：

```
try{
    Class.forName("com.mysql.jdbc.Driver")
}
catch(Exception e){ }
```

（2）与指定的数据库建立连接

如果应用程序与 MySQL 服务器在同台计算机上，那么应用程序与数据库 factory 建立连接的代码如下：

```
String uri= "jdbc:mysql://127.0.0.1/person";
String user="root";
String password="123";
Connection con=DriverManager.getConnection(uri,user,password);
```

如果应用程序与 MySQL 服务器不在同一台计算机上，只需将上述代码中的 127.0.0.1 更改为

MySQL 服务器所在计算机的 IP 地址即可。

其中，root 用户有权访问数据库 person，root 用户的密码是 123。如果 root 用户没有设置密码，那么将上述

    String password="123";

更改为

    String password="";

下面的 Java 应用程序 ConMySQL.java 连接 MySQL 数据库 person、向 message 表插入记录、查询记录（效果如图 12-13 所示）。

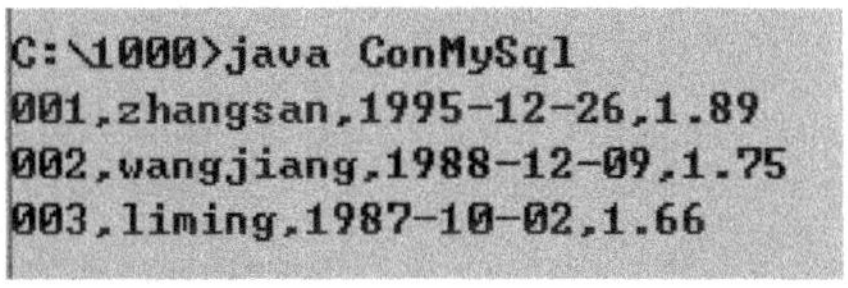

ConMySQL.jave

```
import java.sql.*;
public class ConMySQL{
    public static void main(String args[]){
        Connection con;
        Statement sql;
        ResultSet rs;
        try{
            Class.forName("com.mysql.jdbc.Driver");
        }
        catch(ClassNotFoundException e){
            System.out.println(""+ e);
        }
        try{
            String uri= "jdbc:mysql://127.0.0.1/person",
            user="root",password="123";
            con=DriverManager.getConnection(uri,user,password);
            sql=con.createStatement();
            sql.executeUpdate("INSERT INTO message  VALUES('002','wangjiang','1988-12-9',1.75)");
          sql.executeUpdate("INSERT INTO message VALUES('003','liming','1987-10-2',1.66)");
            sql.executeUpdate("UPDATE message SET height=1.89 WHERE number='001'");
            rs=sql.executeQuery("SELECT * FROM message");
            while(rs.next()){
                String number=rs.getString(1);
                String name=rs.getString(2);
                Date birth=rs.getDate(3);
                double height=rs.getDouble(4);
                System.out.println(number+ ","+ name+ ","+ birth+ ","+ height);
            }
            con.close();
        }
        catch(Exception e){
            System.out.println(e);
        }
    }
}
```

图 12-13　连接 MySQL 数据库

第 13 章

Java Applet

本章导读

✿ 知识概述
✿ 实验 1　播放音频
✿ 实验 2　绘制五角星
✿ 实验 3　左手画圆右手画方
✿ 实验 4　图像渐变
✿ 实验 5　读取服务器中的文件
✿ 知识扩展——Java 2D 简介

13.1 知识概述

一个 Java Applet 也由若干个类组成，Java Applet 不再需要 main()方法，但必须有且只有一个类扩展了 Applet 类，即它是 Applet 类的子类，我们把这个类叫做这个 Java Applet 的主类。Java Applet 的主类必须是 public 的。Java Applet 属于嵌入到浏览器环境中的程序，必须由浏览器中的 JVM 负责执行。当 Java Applet 编译通过之后，必须编写一个超文本文件（含有 applet 标记的 Web 页）告诉浏览器来运行这个 Java Applet。假设 Applet 主类的名字是 Boy，下面是一个简单的 HTML 文件 like.html：

        <Applet code=Boy.class height=180 width=300>
        </Applet>

like.html 文件告诉浏览器运行主类是 Boy 的 Java Applet。

网页的最终目的是让其他客户通过网络来访问，下载到客户机执行，可以用 Web 发布管理器（Windows 98 安装盘可以安装个人 Web 管理器）或 IIS 将含有 Java Applet 网页所在的目录设成 Web 共享。假如将 like.html 所在的文件夹 C:\1000 设为 Web 共享目录，共享的名称是 hello，那么其他用户就可以在其浏览器的地址栏中输入服务器的 IP 地址、共享 Web 目录和该目录中的 HTML 文件（http://192.168.0.100/hello/like.html），来下载或执行含有 Java Applet 程序的网页。也就是说，Java Applet 的字节码文件会下载到客户机，由客户机的浏览器负责运行。

浏览器内置的 JVM 创建负责创建主类的对象，该对象立刻调用 init()方法完成必要的初始化工作。初始化的主要任务是创建所需要的对象、设置初始状态、装载图像、设置参数等。该对象然后自动调用 start()方法。在程序的执行过程中，init()方法只被调用一次。但 start()方法将多次被自动调用。除了进入执行过程时调用方法 start()外，当用户从 Java Applet 所在的 Web 页面转到其他页面然后又返回时，start()将再次被调用，但不再调用 init()方法。当浏览器离开 Java Applet 所在的页面转到其他页面时，主类创建的对象将调用 stop()方法。如果浏览器又回到此页，则 start()方法又被调用。在 Java Applet 的生命周期中，start()方法和 stop()方法可能被调用多次，但 init()方法只被调用一次。当用户关闭浏览器结束浏览时，主类创建的对象自动执行 destroy()方法，结束 Java Applet 的生命。

paint(Graphics g)方法可以使一个 Java Applet 在容器上显示某些信息，如文字、色彩、背景或图像等，在 Java Applet 的生命周期内可以多次被调用。例如，当 Java Applet 被其他页面遮挡，然后又重新放到最前面、改变浏览器窗口的大小及 Java Applet 本身需要显示信息时，主类创建的对象都会自动调用 paint()方法。

13.2 实验练习

13.2.1 播放音频

1. 实验目的

本实验的目的是掌握在 Java Applet 中播放音频的方法。

2. 实验要求

通过 Java Applet 播放多个音频文件，要求用 Choice 类创建一个选择控件组件，将声音文件放在控件的选择列表中，当客户选择列表中一个项目后，就启动一个创建音频对象的线程。另外，含有 Java Applet 的超文本中使用若干个<Param...>标志，把值传递到 Java Applet 中。下面是为运行 Java Applet 程序的 HTML 文件：

```
<Applet   code=PlayAudioClip.class   width=200   height=200>
<Param name="1"    value ="祝酒歌:1.au">
<Param name="2"    value ="云的思念:2.au">
<Param name="3"    value ="祝你平安:3.au">
<Param name="4"    value="难忘今宵:4.au">
<Param name="总数"    value="4">
</Applet>
```

3．运行效果示例

运行效果如图 13-1 所示。

4．程序模板

按模板要求，将【代码】替换为 Java 程序代码。

图 13-1 播放音频

PlayAudioClip.java

```
import java.applet.*;
import java.awt.*;
import java.awt.event.*;
public class PlayAudioClip extends Applet implements ActionListener,Runnable,ItemListener{
    AudioClip clip;
    Choice choice;
    TextField text;
    Thread thread;
    String item=null;
    Button button_play,button_loop,button_stop;
    public void init(){
       choice=new Choice();
       thread=new Thread(this);
       int N=Integer.parseInt(getParameter("总数"));
       for(int i=1;i<=N;i++){
          choice.add(getParameter(String.valueOf(i)));
       }
       button_play=new Button("开始播放");
       button_loop=new Button("循环播放");
       button_stop=new Button("停止播放");
       text=new TextField(12);
       button_play.addActionListener(this);
       button_stop.addActionListener(this);
       button_loop.addActionListener(this);
       choice.addItemListener(this);
       add(choice);
       add(button_play);
       add(button_loop);
       add(button_stop);
```

```
            add(text);
            button_play.setEnabled(false);
            button_loop.setEnabled(false);
        }
        public void itemStateChanged(ItemEvent e){
            item=choice.getSelectedItem();
            int index=item.indexOf(":");
            item=item.substring(index+1).trim();
            if(!(thread.isAlive())){
                thread=new Thread(this);
            }
            try{
                thread.start();
            }
            catch(Exception exp){
                text.setText("正在下载音频文件");
            }
        }
        public void stop(){
            【代码1】                          //clip 停止播放
        }
        public void actionPerformed(ActionEvent e){
            if(e.getSource()==button_play){
                【代码2】                      //clip 开始播放，但不循环播放
            }
            else if(e.getSource()==button_loop){
                【代码3】                      //clip 开始播放，并且循环播放
            }
            else if(e.getSource()==button_stop){
                【代码4】                      //clip 停止播放
                button_play.setEnabled(false);
                button_loop.setEnabled(false);
            }
        }
        public void run(){
            clip=getAudioClip(getCodeBase(),item);
            text.setText("请稍等...");
            if(clip!=null){
                button_play.setEnabled(true);
                button_loop.setEnabled(true);
                text.setText("您可以播放了");
            }
        }
    }
```

5. **实验指导与检查**

⊙ 用 Java 可以编写播放 AU、AIFF、WAV、MIDI 和 RM 格式的音频。AU 格式是 Java 早期

唯一支持的音频格式。
- 可以在超文本中使用若干<Param...>标志把值传递到 Java Applet 中,这样就实现了动态地向程序传递信息,不必重新编译程序,便于程序的维护和使用。
- 上述 Java Applet 需要播放的音频文件必须与 Java Applet 在同一文件目录中。
- 可以使用操作系统提供的搜索功能查找 AU 或 WAV 格式的音频文件。
- 向实验指导教师演示程序的运行效果。

6．实验报告

实验报告的格式如下（可要求学生填写并由实验指导教师签字）：

学号：_____ 班级：_____ 姓名：_____ 时间：_____

实 验 内 容	回　答	教 师 评 语
在 HTML 文件中再增加<Param...>标志,使得 Java Applet 能播放扩展名是 WAV 格式的音频文件		

13.2.2 绘制五角星

图 13-2　绘制五角星

1．实验目的
本实验的目的是掌握使用 Graphics 对象绘制多边形的方法。

2．实验要求
绘制一个五角星。

3．运行效果示例
运行效果如图 13-2 所示。

4．程序模板
按模板要求,将【代码】替换为 Java 程序代码。

Star.java

```
import java.applet.*;
import java.awt.*;
public class Star extends Applet{
   int pointX[]=new int[5],
   pointY[]=new int[5];
   public void paint(Graphics g){
      g.translate(100,100);
      pointX[0]=0;
      pointY[0]=-80;
      double arcAngle=(72*Math.PI)/180;
      for(int i=1;i<5;i++){
         pointX[i]=(int)(pointX[i-1]*Math.cos(arcAngle)-pointY[i-1]*Math.sin(arcAngle));
         pointY[i]=(int)(pointY[i-1]*Math.cos(arcAngle)+pointX[i-1]*Math.sin(arcAngle));
      }
      g.setColor(Color.red);
      int starX[]={pointX[0],pointX[2],pointX[4],pointX[1],pointX[3],pointX[0]};
```

```
            int starY[]={pointY[0],pointY[2],pointY[4],pointY[1],pointY[3],pointY[0]};
            【代码1】                        // g 绘制由数组 starX 和 starY 对应点组成的多边形
        }
    }
```

5．实验指导与检查

⊙ Graphics 对象调用 translate(int x,int y)方法可进行坐标变换。

⊙ 向实验指导教师演示程序的运行效果。

6．实验报告

实验报告的格式如下（可要求学生填写并由实验指导教师签字）：

学号：_____   班级：_____   姓名：_____   时间：_____

实 验 内 容	回　　答	教 师 评 语
绘制一个正六边形		

13.2.3　左手画圆右手画方

1．实验目的

本实验的目的是掌握使用多线程绘制图形的方法。

2．实验要求

在 Java Applet 中有两个线程：一个线程模拟左手画圆，另一个线程模拟右手画正方形，要求圆是正方形的内切圆。观察哪个线程首先绘制完毕。

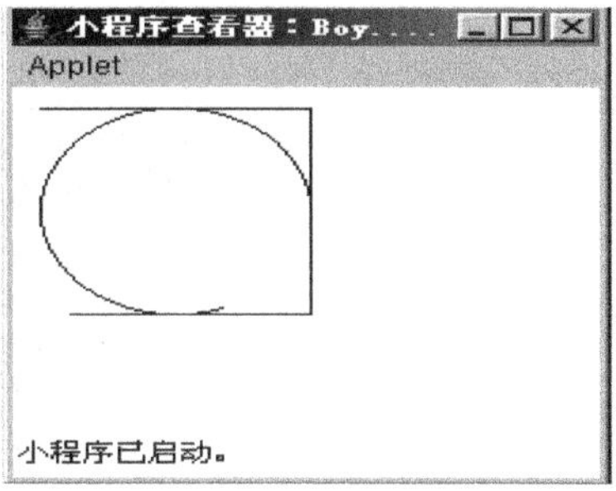

3．运行效果示例

运行效果如图 13-3 所示。

图 13-3　双线程绘制图形

4．程序模板

按模板要求，将【代码】替换为程序代码。

Boy.java

```
    import java.applet.*;
    import java.awt.*;
    public class Boy extends Applet implements Runnable{
        Thread left,right;
        Graphics gLeft,gRight;
        int n=0,m=10;
        public void init(){
            left=new Thread(this);
            right=new Thread(this);
            gLeft=getGraphics();
            gRight=getGraphics();
        }
        public void start(){
            left.start();
```

```
                right.start();
        }
        public void run(){
            while(true){
                if(Thread.currentThread()==left){
                    n++;
                    if(n>=360){
                        n=1;
                    }
                    【代码1】            // gLeft 绘制圆弧,绘制参数为(10,10,100,100,0,n)
                    try{
                        Thread.sleep(100);
                    }
                    catch(InterruptedException e){ }
                }
                else if(Thread.currentThread()==right){
                    m++;
                    if(m<=110){
                        【代码2】        // gRight 绘制直线,绘制参数为(10,10,m,10)
                    }
                    else if(m>110&&m<=210) {
                        【代码3】        // gRight 绘制直线,绘制参数为(110,10,110,10+ m-110)
                    }
                    else if(m>210&&m<=310){
                        【代码4】        // gRight 绘制直线,绘制参数为(110,110,110-(m-210),110)
                    }
                    else if(m>310&&m<=410) {
                        【代码5】        // gRight 绘制直线,绘制参数为(10,110,10,110-(m-310))
                    }
                    else if(m>410){
                        m=10;
                    }
                    try{
                        Thread.sleep(100);
                    }
                    catch(InterruptedException e){ }
                }
            }
        }
    }
```

5. 实验指导与检查

- Java Applet 调用 getGraphics()方法可以返回一个 Graphics 对象。
- 向实验指导教师演示程序的运行效果。

6. 实验报告

实验报告的格式如下（可要求学生填写并由实验指导教师签字）：

实 验 内 容	回　　答	教师评语
编写双线程绘制图形的 Java Applet，一个线程顺时针绘制圆，另一个线程逆时针绘制圆		

13.2.4　图像渐变

1．实验目的

本实验的目的是掌握在 Java Applet 中绘制图像的方法。

2．实验要求

在 Java Applet 中有一个线程，该线程逐渐放大一幅图像。

3．运行效果示例

运行效果如图 13-4 所示。

4．程序模板

按模板要求，将【代码】替换为 Java 程序代码。

Boy.java

图 13-4　图像逐渐变化大小

```
import java.applet.*;
import java.awt.*;
public class Boy extends Applet implements Runnable{
    Thread drawImage;
    Image image;
    Graphics g;
    int w=10,h=10;
    public void init(){
        image=getImage(getCodeBase(),"a.gif");
        drawImage=new Thread(this);
        g=getGraphics();
    }
    public void start(){
        drawImage.start();
    }
    public void run(){
        while(true){
            g.clearRect(10,10,w,h);
            w++;
            h++;
            if(w>=300){
                w=10;
                h=10;
            }
            【代码1】               // g 在矩形(10,10,w,h)中绘制图像 image
            try{
```

```
                    Thread.sleep(100);
                }
                catch(InterruptedException e){ }
            }
        }
    }
```

5．实验指导与检查

- Java Applet 调用 getGraphics()方法可以返回一个 Graphics 对象。
- 向实验指导教师演示程序的运行效果。

6．实验报告

实验报告的格式如下（可要求学生填写并由实验指导教师签字）：

学号：_____　　班级：_____　　姓名：_____　　时间：_____

实 验 内 容	回　　答	教 师 评 语
编写双线程绘制图像的 Java Applet，一个线程逐渐放大一幅图像，另一个线程逐渐缩小一幅图像		

13.2.5　读取服务器端文件

1．实验目的

本实验的目的是掌握在 Java Applet 中使用 URL 对象的方法。

2．实验要求

首先根据 Java Applet 的字节码驻留在服务器端地址创建一个 URL 对象，然后让 URL 对象返回输入流，通过该输入流读取 Java Applet 所在目录下的文件，文件的名字通过 HTML 文件传递到 Java Applet 中。以下是含有 Java Applet 的 Web 文件：

```
<Applet code=ReadFile.class  width=200  height=200>
<Param name="1"   value ="Example.java">
<Param name="2"   value ="a.txt">
<Param name="3"   value ="E.java">
<Param name="总数"  value="3">
<WApplet>
```

3．运行效果示例

运行效果如图 13-5 所示。

4．程序模板

按模板要求，将【代码】替换为 Java 程序代码。

ReadFile.java
```
import java.applet.*;
import java.awt.*;
import java.awt.event.*;
import java.net.*;
```

图 13-5 读取文件

```
import java.io.*;
public class ReadFile extends Applet implements ActionListener,Runnable{
    File file;
    Choice choice;
    TextArea text;
    Thread thread;
    String item=null;
    Button button;
    URL url;
    public void init(){
        choice=new Choice();
        thread=new Thread(this);
        int N=Integer.parseInt(getParameter("总数"));
        for(int i=1;i<=N;i++){
            choice.add(getParameter(String.valueOf(i)));
        }
        button=new Button("开始读取");
        text=new TextArea(12,40);
        button.addActionListener(this);
        add(choice);
        add(button);
        add(text);
    }
    public void actionPerformed(ActionEvent e){
        text.setText(null);
        item=choice.getSelectedItem();
        item=item.trim();
        if(!(thread.isAlive())){
            thread=new Thread(this);
        }
        try{
```

```
            thread.start();
        }
        catch(Exception exp){ }
    }
    public void run(){
        try{
            url=new URL(getCodeBase(),item);
            InputStream in=【代码1】                    // url 返回输入流
            int n=-1;
            byte b[]=new byte[100];
            while((n=in.read(b))!=-1){
                String str=new String(b,0,n);
                text.append(str);
            }
        }
        catch(Exception ee){ }
    }
}
```

5. 实验指导与检查

⦿ Applet 类的方法

      public URL getCodeBase()

返回一个 URL 对象，该对象包含 Java Applet 所在的目录。假如 Java Applet 所在的目录是 java，那么返回的 URL 对象是 "http://192.168.0.1.200/java"。Applet 还有一个方法

      public URL getDocumentBase()

该方法返回一个 URL 对象，该对象是嵌入 Java Applet 的网页的 URL，如返回的 URL 对象含有的信息 "http://192.168.0.1.200/java/ReadFile.html"。

⦿ 向实验指导教师演示程序的运行效果。

6. 实验报告

实验报告的格式如下（可要求学生填写并由实验指导教师签字）：

学号：_____ 班级：_____ 姓名：_____ 时间：_____

实 验 内 容	回　　答	教师评语
将程序中的 AWT 组件换成 Swing 组件		

13.3　知识扩展——Java 2D 简介

Java 1.2 给出了一个新类 Graphics2D，它是 Graphics 类的子类。Graphics2D 对象把直线、圆等作为一个对象来绘制。也就是说，如果想用一个 Graphics2D 类型的"画笔"来画一个圆，就必须先创建一个圆的对象，仍需使用 paint(Graphics g)方法来绘制。Java 允许将 Graphics 对象强制转化为 Graphics2D 对象。

1．绘制直线

使用 java.awt.geom 包中的 Line2D 的子类 Line2D.Double 创建一个直线对象，Double(double x1, double y1, double x2, double y2)方法创建一个(x1,y1)到 (x2,y2)的 Line2D 对象。例如：

    Line2D line=new Line2D.Double(2,2,300,300);

然后 Graphics2D 对象 g 调用 draw()方法绘制该直线：

    g.draw(line);

2．绘制矩形

使用 java.awt.geom 包中的 Rectangle2D.Double 类来创建一个矩形对象，如语句

    Rectangle2D rect=new Rectangle2D. Double(50,50,300,60.897);

创建了一个左上角坐标是(50,50)、宽是 300、高是 60.987 的一个矩形对象。然后，Graphics2D 对象 g 调用 draw()方法或 fill()方法绘制矩形或填充矩形：

    g.draw(rect)
    g.fill(rect)

3．绘制圆角矩形

使用 java.awt.geom 包中的 RoundRectangle2D.Double 类来创建一个圆角矩形对象，如语句

    RoundRectangle2D rect_round=new RoundRectangle2D.Double(50,50,300,40,8,5);

创建了一个左上角坐标是(50,50)、宽是 300、高是 40、圆角的长轴和短轴分别为 8 和 5 的圆角矩形对象。然后，Graphics2D 对象 g 调用 draw()方法或 fill()方法绘制或填充该图形。

4．绘制椭圆

使用 java.awt.geom 包中的 Ellipse2D. Double 类来创建一个椭圆对象，如语句

    Ellipse2D ellipse=new Ellipse2D. Double (50,30,300,60);

创建了一个外接矩形的左上角坐标是(50,30)、宽是 300、高是 60 的椭圆对象。然后，Graphics2D 对象 g 调用 draw()方法或 fill()方法绘制或填充该图形。

5．绘制圆弧

使用 java.awt.geom 包中的 Arc2D. Double 类创建一个圆弧对象，如语句

    Arc2D arc=new Arc2D. Double (50,30,300,70,0,100,Arc.PIE);

创建了一个外接矩形的左上角坐标是(50,30)、宽是 300、高是 70、起始角是 0°、终止角是 100°的饼弧对象。最后一个参数取值 Arc.OPEN、Arc.CHORD 或 Arc.PIE，决定弧是开弧、弓弧或饼弧。然后，Graphics2D 对象 g 调用 draw()方法或 fill()方法绘制或填充该图形。

6．绘制二次曲线

二次曲线可用二阶多项式

$$y(x)=ax^2+bx+c$$

来表示。一条二次曲线需要三个点来确定。

使用 java.awt.geom 包中的 QuadCurve2D.Double 类来创建一个二次曲线：

    QuadCurve2D curve=new QuadCurve2D.Double (50,30,10,10,50,100);

上述代码片段使用端点(50, 30)和(50, 100)及控制点(10, 10)创建了一条二次曲线。

7．绘制三次曲线

三次曲线可用三阶多项式

$$y(x)=ax^3+bx^2+cx+d$$

来表示。一条三次曲线需要 4 个点来确定该曲线。

使用 java.awt.geom 包中的 QuadCurve2D.Double 类来创建一个二次曲线：

CubicCurve2D curve=new QuadCurve2D.Double (50,30,10,10,100,100,50,100);

上述语句使用端点(50,30)和(50,100)及控制点(10,10)和(100,100)创建了一条三次曲线。然后，Graphics2D 对象 g 调用 draw()方法或 fill()方法绘制或填充该图形。

8．控制图形线条的粗细

Graphics 类创建的"画笔"的粗细是默认的，用户不能改变它。在 Java 2D 中可以改变画笔的粗细，使用 BasicStroke 类创建一个供画笔选择线条粗细的对象。BasicStroke 类的一个常用的构造方法如下：

BasicStroke (float width, int cap, int join)

其中，width 参数决定画笔线条的粗细，默认值是 1；cap 参数决定线条两端的形状，对于与周围不再有连接的直线或曲线是有用的，其取值是 BasicStroke.CAP_BUTT、BisicStroke.CAP_ROUND 或 BisicStroke.CAP_SQUARE；参数 join 决定线条中的角应如何处理，其取值是 BasicStroke.JOIN_BEVEL、BisicStroke.JOIN_MITER 或 BisicStroke.JOIN_ROUND。

然后，Graphics2D 对象调用方法 setStroke(BasicStroke a)设置线条形状。

9．旋转图形

我们有时需要平移、缩放或旋转一个图形，可以使用 AffineTransform 类来实现对图形的这些操作。

第一步：首先使用 AffineTransform 类创建一个对象：

AffineTransform trans=new AffineTransform();

对象 trans 常用下列 3 种方法实现对图形的操作：

- translate(double a,double b) ——将图形在 X 轴方向移动 a 单位像素，Y 轴方向移动 b 像素单位。x 是正值时向右移动，是负值时向左移动；y 是正值时向下移动，是负值时向上移动。
- scale(double a,double b) ——将图形在 X 轴方向缩放 a 倍，在 Y 轴方向缩放 b 倍。
- rotate(double number,double x,double y) ——将图形沿顺时针或逆时针以(x,y)为轴点旋转 number 弧度。

第二步：进行需要的变换。例如，要把一个矩形绕(100,100)点顺时针旋转 60°，那么就要做好准备：

trans.rotate(60.0*3.1415927/180,100,100);

第三步：把 Graphics2D 对象如 g_2d 设置为具有 trans 这种功能的"画笔"：

g_2d.setTransform(trans);

假如 rect 是一个矩形对象，那么 g_2d.draw(rect)画的就是旋转后的矩形的样子。

注意：不要把第二步和第三步颠倒。

下面的 Boy.java 绘制了一个太极图、一条抛物线，并旋转了一个椭圆，如图 13-6 所示。

Boy.jave
```
import java.awt.*;
import java.applet.*;
import java.awt.geom.*;
public class Boy extends Applet{
    public void init(){
        setBackground(Color.yellow);
    }
    public void paint(Graphics g){
        Graphics2D g_2d=(Graphics2D)g;
        g_2d.setColor(Color.black);
        Arc2D arc=new Arc2D.Double(0,0,100,100,-90,-180,Arc2D.PIE);
        g_2d.fill(arc);
        g_2d.setColor(Color.white);
        arc=new Arc2D.Double(0,0,100,100,-90,180,Arc2D.PIE);
        g_2d.fill(arc);
        arc=new Arc2D.Double(25,0,50,50,-90,-180,Arc2D.PIE);
        g_2d.fill(arc);
        g_2d.setColor(Color.black);
        Ellipse2D ellipse1=new Ellipse2D.Double(40,15,20,20);
        g_2d.fill(ellipse1);
        arc=new Arc2D.Double(25,50,50,50,90,-180,Arc2D.PIE);
        g_2d.fill(arc);
        g_2d.setColor(Color.white);
        ellipse1=new Ellipse2D.Double(40,65,20,20);
        g_2d.fill(ellipse1);
        g_2d.setColor(Color.blue);
        QuadCurve2D curve=new QuadCurve2D.Double(0,120,50,280,100,120);
        g_2d.draw(curve);
        Ellipse2D ellipse2=new Ellipse2D.Double(150,80,120,50);
        BasicStroke bs=new BasicStroke(6,BasicStroke.CAP_BUTT,BasicStroke.JOIN_BEVEL);
        g_2d.setStroke(bs);
        AffineTransform trans=new AffineTransform();
        for(int i=1;i<=12;i++){
            trans.rotate(30.0*Math.PI/180,210,105);
            g_2d.setTransform(trans);
            g_2d.draw(ellipse2);
        }
    }
}
```

图 13-6 填充和旋转

第14章

综合实验——走迷宫

本章导读

- ✪ 设计要求
- ✪ 总体设计
- ✪ 详细设计
- ✪ 代码编写与调试
- ✪ 软件发布
- ✪ 实验后的练习

14.1 设计要求

设计 GUI 界面的走迷宫游戏，游戏的结果是让走迷宫者从迷宫的入口处走到迷宫的出口处。具体要求如下：

（1）程序根据文本文件生成迷宫，这些文本文件称为迷宫文件。迷宫文件的扩展名为".maze"，其中的文本内容有着特殊的组织结构：任意两行文本所含有的字符个数必须相同，而且字符只可以是"*"、"#"、"0"或"1"。迷宫文件中的"*"和"#"分别代表迷宫的入口和出口，"0"和"1"分别代表迷宫中的"路"和"墙"。例如，名为"简单迷宫.maze"的迷宫文件所生成的迷宫如图14-1 所示。

（2）用户可以通过界面上提供的菜单选项，选择"迷宫文件"生成对应的"迷宫"。

（3）用户可以通过界面上提供的菜单选项，选择迷宫中的"墙"和"路"的外观图像。

（4）用户可以随时单击界面上提供的按钮，重新开始走"迷宫"。

14.2 总体设计

在设计走迷宫时编写如下 6 个 Java 源文件：MazeWindow.java，Maze.java，WallOrRoad.java，MazePoint.java，PersonInMaze.java 和 HandleMove.java。除了需要编写的上述 6 个 Java 源文件所给出的类外，还需要 Java 系统提供一些重要的类，如 JMenu 类和 File 类等。走迷宫所用到的一些重要的类及其组合关系如图 14-2 所示。

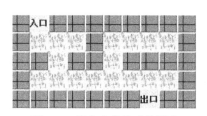

图14-1　迷宫文件生成的迷宫　　　　图14-2　类之间的组合关系

1. 迷宫文件

走迷宫通过使用文本文件生成迷宫，以便灵活、有效地设计迷宫。允许用户自己编写迷宫文件，只要将这些迷宫文件和应用程序存放到同一目录中即可。

2. MazeWindow.java（主类）

MazeWindow 类负责创建走迷宫的主窗口，含有 main()方法，程序从该类开始执行，包括 4 种重要类型的对象：Maze、JMenu、File 和 JButton。我们将在后面的详细设计中阐述 MazeWindow 类的主要成员的作用，MazeWindow 类创建的窗口及其主要成员对象如图 14-3 所示。

3. Maze.java

Maze 类创建的对象是 MazeWindow 类最重要的成员之一，代表迷宫。该类的成员变量中有 4 种重要类型的对象：MazePoint、WallOrRoad、HandleMove 和 PersonInMaze。我们将在后面的详细设计中阐述 MazeWindow 类的主要成员的作用。

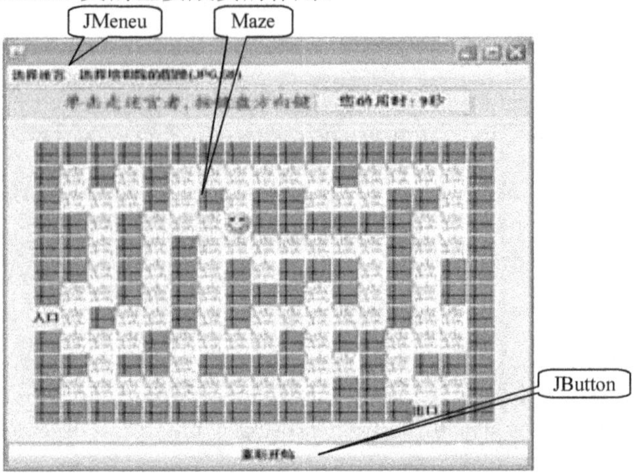

图14-3 MazeWindow窗口及主要的成员对象

4. WallOrRoad.java

WallOrRoad 类是 JPanel 的一个子类，创建的对象是 Maze 类的重要成员之一，用来表示迷宫中的"墙"或"路"。

5. MazePoint.java

MazePoint 类负责创建确定位置的对象，以便程序使用 MazePoint 对象来确定 WallOrRoad 和 PersonInMaze 对象在 Maze 对象中的位置，即确定"墙"和"路"，以及"走迷宫者"在迷宫中的位置。

6. PersonInMaze.java

PersonInMaze 类所创建的对象代表"走迷宫者"。

7. HandleMove.java

HandleMove 类所创建的对象负责处理键盘事件。

14.3 详细设计

14.3.1 编写迷宫文件

为了便于管理迷宫文件，走迷宫程序将迷宫文件的扩展名规定为".maze"。使用文本编辑器，如"记事本"编辑迷宫文件，在保存迷宫文件时，必须将"保存类型"选择为"所有文件"，将"编码"选择为"ANSI"。如果在保存迷宫文件时，系统总是给文件名末尾加上".txt"扩展名，那么在保存迷宫文件时可以将迷宫文件的名字用双引号扩起，如图 14-4 所示。

图14-4 迷宫文件的保存

可以编写若干个迷宫文件。迷宫文件中任意两行中的字符个数相同，字符可以是"*"、"#"、

"0"或"1",除此之外,不允许含有其他可见字符(字符之间不要有空格)。迷宫文件中的"*"和"#"分别代表迷宫的入口和出口,"0"和"1"分别代表迷宫中的"路"和"墙"。我们一共编写了 5 个迷宫文件:蜀道迷宫.maze,山水迷宫.maze,代迷宫.maze,味迷宫.maze,宫.maze。其中,"蜀道迷宫.maze"的内容如下:

```
1111111111111111
1010100000100001
1000101011000110l
1101000011111001
1101010000001001
1101010101110101l
1001010111010101l
*010010100000100l
1000100001011000l
110110111100101l1
1000000000011000l
11111111111111#11
```

14.3.2 MazeWindow 类

1. 效果图

MazeWindow 创建的窗口效果如图 14-5 所示。

2. UML 图

MazeWindow 类是 javax.swing 包中 JFrame 的一个子类,并实现了 ActionListener 接口,该类的主要成员变量和方法如图 14-6 所示。

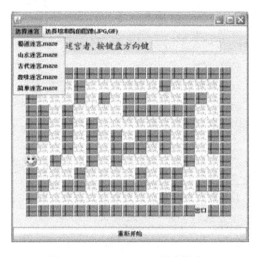

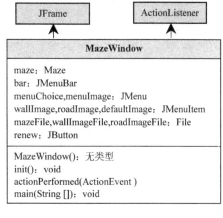

图14-5 MazeWindow创建的窗口　　　　图14-6 Mazewindow类的UML图

(1)成员变量

- maze 是 Maze 类声明的对象,用来刻画迷宫,是 MazeWindow 类中最重要的成员之一。MazeWindow 类根据迷宫文件,即根据该类中的 mazeFile 来创建 maze 对象。
- bar 是 JMenuBar 创建的菜单条,用来添加菜单。bar 被放置在窗口的顶部。

- menuChoice 和 menuImage 是 JMenu 创建的菜单，分别命名为"选择迷宫文件"、"选择墙和路的图像(JPG, GIF)"。menuChoice 和 menuImage 被添加到菜单条 bar 中。
- wallImage、roadImage 和 defaultImage 是 JMenuItem 创建的命令，名字依次为"墙的图像"、"路的图像"和"墙和路的默认图像"，这三个命令被添加到菜单 menuImage 中。wallImage、roadImage 和 defaultImage 都将当前窗口注册为自己的 ActionEvent 事件监视器。
- mazeFile、wallImageFile 和 roadImageFile 是 File 对象。其中，mazeFile 存放迷宫文件的引用，wallImageFile 和 roadImageFile 分别存放迷宫中"墙"和"路"的外观图像文件的引用。
- renew 是 JButton 创建的按钮对象，名为"重新开始"。renew 将当前窗口注册为自己的 ActionEvent 事件监视器。

（2）方法
- MazeWindow()是构造方法，负责完成窗口的初始化操作，其重要的操作之一是：读取当前目录中扩展名为".maze"的迷宫文件，根据迷宫文件的名字创建 JMenuItem 的命令，并将创建的每个 JMenuItem 命令添加到 menuChoice 菜单中，然后将当前窗口注册为每个命令的 ActionEvent 事件监视器。另外，该构造方法在执行过程中，将 mazeFile 初始化为 menuChoice 菜单的第一个命令所代表的迷宫文件，并根据该迷宫文件创建 MazeWindow 窗口中的 maze 对象。该构造方法还对 wallImageFile 和 roadImageFile 进行初始化，并指定迷宫中"墙"和"路"所使用的默认图像。
- init()方法根据迷宫文件初始化 maze 对象，完成必要的初始化操作。
- actionPerformed(ActionEvent)方法是 MazeWindow 类实现的 ActionListener 接口中的方法。MazeWindow 类创建的窗口是 menuChoice 菜单和 menuImage 菜单中的命令及 renew 按钮上的 ActionEvent 事件监视器。当用户选择某个命令或单击按钮时，窗口将执行 actionPerformed(ActionEvent)方法进行相应的操作。如果用户选择 menuChoice 菜单中的某个命令，即用户选择了一个迷宫文件，actionPerformed(ActionEvent)方法执行的操作就是改变 mazeFile 引用的迷宫文件，然后调用 init()方法。如果用户选择 menuImage 菜单中的 wallImage 或 roadImage 命令，actionPerformed(ActionEvent)方法执行的操作就是分别改变 wallImageFile 和 roadImageFile 引用的图像文件，从而让 maze 对象改变"墙"或"路"的图像；如果用户选择 menuImage 菜单的 defaultImage 命令，actionPerformed(ActionEvent)方法执行的操作就是将 wallImageFile 和 roadImageFile 引用的图像文件恢复到默认设置，让 maze 对象恢复"墙"或"路"的默认图像。单击 renew 按钮时，actionPerformed(ActionEvent)方法进行的操作是保持当前的 mazeFile 所引用的迷宫文件，并执行 init()方法。
- main()方法是程序运行的入口方法。

3. 代码（MazeWindow.java）

```
import javax.swing.*;
import java.awt.*;
import java.awt.event.*;
import java.io.*;
import javax.swing.filechooser.*;
public class MazeWindow extends JFrame implements ActionListener{
    Maze maze;
```

```java
JMenuBar bar;
JMenu menuChoice,menuImage;
JMenuItem wallImage,roadImage,defaultImage;
File mazeFile,wallImageFile,roadImageFile;
JButton renew;
MazeWindow(){
    wallImageFile=new File("wall.jpg");
    roadImageFile=new File("road.jpg");
    bar=new JMenuBar();
    menuChoice=new JMenu("选择迷宫");
    File dir=new File(".");
    File file[]=dir.listFiles(new FilenameFilter(){
                                 public boolean accept(File dir,String name){
                                     return name.endsWith("maze");
                                 }
                              });
    for(int i=0;i<file.length;i++){
        JMenuItem item=new JMenuItem(file[i].getName());
        item.addActionListener(this);
        menuChoice.add(item);
    }
    mazeFile=new File(file[0].getName());
    init();
    menuImage=new JMenu("选择墙和路的图像(JPG, GIF)");
    wallImage=new JMenuItem("墙的图像");
    roadImage=new JMenuItem("路的图像");
    defaultImage=new JMenuItem("墙和路的默认图像");
    menuImage.add(wallImage);
    menuImage.add(roadImage);
    menuImage.add(defaultImage);
    bar.add(menuChoice);
    bar.add(menuImage);
    setJMenuBar(bar);
    wallImage.addActionListener(this);
    roadImage.addActionListener(this);
    defaultImage.addActionListener(this);
    renew=new JButton("重新开始");
    renew.addActionListener(this);
    add(maze,BorderLayout.CENTER);
    add(renew,BorderLayout.SOUTH);
    setVisible(true);
    setBounds(60,60,510,480);
    validate();
    setDefaultCloseOperation(JFrame.EXIT_ON_CLOSE);
}
public void init(){
    if(maze!=null){
```

```java
            remove(maze);
            remove(maze.getHandleMove());
        }
        maze=new Maze();
        maze.setWallImage(wallImageFile);
        maze.setRoadImage(roadImageFile);
        maze.setMazeFile(mazeFile);
        add(maze,BorderLayout.CENTER);
        add(maze.getHandleMove(),BorderLayout.NORTH);
        validate();
    }
    public void actionPerformed(ActionEvent e){
        if(e.getSource()==roadImage){
            JFileChooser chooser=new JFileChooser();
            FileNameExtensionFilter filter = new FileNameExtensionFilter("JPG & GIF Images", "jpg", "gif");
            chooser.setFileFilter(filter);
            int state=chooser.showOpenDialog(null);
            File file=chooser.getSelectedFile();
            if(file!=null&&state==JFileChooser.APPROVE_OPTION){
                roadImageFile=file;
                maze.setRoadImage(roadImageFile);
            }
        }
        else if(e.getSource()==wallImage){
            JFileChooser chooser=new JFileChooser();
            FileNameExtensionFilter filter = new FileNameExtensionFilter("JPG & GIF Images", "jpg", "gif");
            chooser.setFileFilter(filter);
            int state=chooser.showOpenDialog(null);
            File file=chooser.getSelectedFile();
            if(file!=null&&state==JFileChooser.APPROVE_OPTION){
                wallImageFile=file;
                maze.setWallImage(wallImageFile);
            }
        }
        else if(e.getSource()==defaultImage){
            wallImageFile=new File("wall.jpg");
            roadImageFile=new File("road.jpg");
            maze.setWallImage(wallImageFile);
            maze.setRoadImage(roadImageFile);
        }
        else if(e.getSource()==renew){
            init();
        }
        else{
            JMenuItem item=(JMenuItem)e.getSource();
            mazeFile=new File(item.getText());
            init();
```

            }
        }
        public static void main(String args[]){
            new MazeWindow();
        }
    }

14.3.3 Maze 类

1. 效果图

Maze 根据"蜀道迷宫.maze"迷宫文件（见 14.3.1 节）创建的迷宫效果如图 14-7 所示。

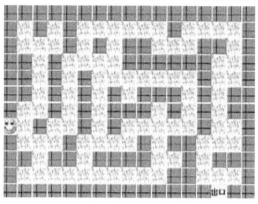

图14-7　Maze创建的对象

2. UML 图

Maze 类是 javax.swing 包中 JLayeredPane 容器的子类，所创建的对象 maze 是 MazeWindow 类中最重要的成员之一，作为一个容器添加到 MazewWindow 窗口的中心。Maze 类的主要成员变量、方法及与 MazeWindow 类之间的组合关系如图 14-8 所示。

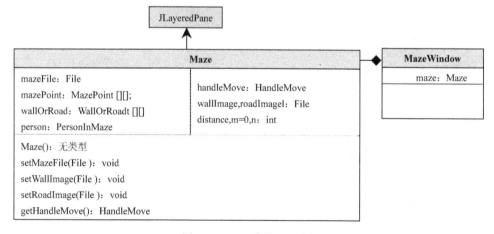

图14-8　Maze类的UML图

（1）成员变量

⊙ mazeFile 是 File 类声明的对象，用来存放迷宫文件的引用。

⊙ mazePoint 是 MazePoint 类型的二维数组，其单元为 MazePoint 类创建的对象，用来确定

"墙"和"路"及"走迷宫者"的位置，即确定 WallOrRoad 和 PersonInMaze 对象在 Maze 对象中的位置。
- wallOrRoad 是 WallOrRoad 类型的二维数组，其单元为 WallOrRoad 类创建的对象，表示迷宫中的"墙"或"路"。
- person 是 PersonInMaze 类创建的对象，表示迷宫中的"走迷宫者"。
- handleMove 是 HandleMove 类创建的对象，负责处理 person 对象上的键盘事件。
- wallImage 和 roadImage 是 File 声明的对象，用来存放绘制"墙"和"路"的外观的图像文件的引用。
- distance、m 和 n 是 int 类型数据。mazePoint 对象根据 distance 的值来进行初始化，以便确定 mazePoint 对象之间的距离。m 和 n 的值是二维数组 mazePoint 的行数和列数。

（2）方法
- Maze()是构造方法，负责创建 Maze 对象。
- Maze 对象调用 setMazeFile(File)方法可根据参数指定的迷宫文件完成必要的初始化，如创建 mazePoint 和 wallOrRoad 数组等。
- Maze 对象调用 setWallImage (File)方法可以设置 wallImage 文件对象。
- Maze 对象调用 setRoadImage(File)方法可以设置 roadImage 文件对象。
- Maze 对象调用 getHandleMove()返回 handleMove 对象。

3. 代码（Maze.java）

```java
import java.awt.*;
import java.awt.event.*;
import javax.swing.*;
import java.io.*;
public class Maze extends JLayeredPane{
    File mazeFile;
    MazePoint[][] mazePoint;
    WallOrRoad[][] wallOrRoad;
    PersonInMaze person;
    HandleMove handleMove;
    File wallImage,roadImage;
    int distance=26,m=0,n=0;
    public Maze(){
        setLayout(null);
        wallImage=new File("wall.jpg");
        roadImage=new File("road.jpg");
        person=new PersonInMaze();
        handleMove=new HandleMove();
        handleMove.initSpendTime();
        person.addKeyListener(handleMove);
        setLayer(person,JLayeredPane.DRAG_LAYER);
    }
    public void setMazeFile(File f){
        mazeFile=f;
        char[][] a;
```

```java
RandomAccessFile in=null;
String lineWord=null;
try{
    in=new RandomAccessFile(mazeFile,"r");
    long length=in.length();
    long position=0;
    in.seek(position);
    while(position<length){
        String str=in.readLine().trim();
        if(str.length()>=n){
            n=str.length();
        }
        position=in.getFilePointer();
        m++;
    }
    a=new char[m][n];
    position=0;
    in.seek(position);
    m=0;
    while(position<length){
        String str=in.readLine();
        a[m]=str.toCharArray();
        position=in.getFilePointer();
        m++;
    }
    in.close();
    wallOrRoad=new WallOrRoad[m][n];
    for(int i=0;i<m;i++){
        for(int j=0;j<n;j++){
            wallOrRoad[i][j]=new WallOrRoad();
            if(a[i][j]=='1'){
                wallOrRoad[i][j].setIsWall(true);
                wallOrRoad[i][j].setWallImage(wallImage);
                wallOrRoad[i][j].repaint();
            }
            else if(a[i][j]=='0'){
                wallOrRoad[i][j].setIsRoad(true);
                wallOrRoad[i][j].setRoadImage(roadImage);
                wallOrRoad[i][j].repaint();
            }
            else if(a[i][j]=='*'){
                wallOrRoad[i][j].setIsEnter(true);
                wallOrRoad[i][j].setIsRoad(true);
                wallOrRoad[i][j].repaint();
            }
            else if(a[i][j]=='#'){
                wallOrRoad[i][j].setIsOut(true);
```

```java
                    wallOrRoad[i][j].setIsRoad(true);
                    wallOrRoad[i][j].repaint();
                }
            }
        }
        mazePoint=new MazePoint[m][n];
        int Hspace=distance,Vspace=distance;
        for(int i=0;i<m;i++){
            for(int j=0;j<n;j++){
                mazePoint[i][j]=new MazePoint(Hspace,Vspace);
                Hspace=Hspace+distance;
            }
            Hspace=distance;
            Vspace=Vspace+distance;
        }
        for(int i=0;i<m;i++){
            for(int j=0;j<n;j++){
                add(wallOrRoad[i][j]);
                wallOrRoad[i][j].setSize(distance,distance);
                wallOrRoad[i][j].
                setLocation(mazePoint[i][j].getX(),mazePoint[i][j].getY());
                wallOrRoad[i][j].setAtMazePoint(mazePoint[i][j]);
                mazePoint[i][j].setWallOrRoad(wallOrRoad[i][j]);
                mazePoint[i][j].setIsWallOrRoad(true);
                if(wallOrRoad[i][j].getIsEnter()){
                    person.setAtMazePoint(mazePoint[i][j]);
                    add(person);
                    person.setSize(distance,distance);
                    person.
                    setLocation(mazePoint[i][j].getX(),mazePoint[i][j].getY());
                    person.requestFocus();
                    person.repaint();
                }
            }
        }
        handleMove.setMazePoint(mazePoint);
    }
    catch(IOException exp){
        JButton mess=new JButton("无效的迷宫文件");
        add(mess);
        mess.setBounds(30,30,100,100);
        mess.setFont(new Font("宋体",Font.BOLD,30));
        System.out.println(exp+ "mess");
    }
}
public void setWallImage(File f){
    wallImage=f;
```

```
                for(int i=0;i<m;i++){
                    for(int j=0;j<n;j++){
                        if(wallOrRoad[i][j].getIsWall())
                            wallOrRoad[i][j].setWallImage(wallImage);
                            wallOrRoad[i][j].repaint();
                        }
                    }
                }
            }
            public void setRoadImage(File f){
                roadImage=f;
                for(int i=0;i<m;i++){
                    for(int j=0;j<n;j++){
                        if(wallOrRoad[i][j].getIsRoad())
                            wallOrRoad[i][j].setRoadImage(roadImage);
                            wallOrRoad[i][j].repaint();
                        }
                    }
                }
            }
            public HandleMove getHandleMove(){
                return handleMove;
            }
        }
```

14.3.4 WallOrRoad 类

1. 效果图

WallOrRoad 创建的对象效果如图 14-9 所示。

图 14-9 WallOrRoad 创建的 2 个对象

2. UML 图

WallOrRoad 类是 javax.swing 包中 JPanel 容器的一个子类,创建的对象是二维数组 wallOrRoad 的单元中的对象。WallOrRoad 型数组 wallOrRoad 是 Maze 类的重要成员之一。WallOrRoad 类的主要成员变量和方法及与 Maze 类之间的关系如图 14-10 所示。

(1) 成员变量

- isRoad、isWall、isEnter 和 isOut 是 boolean 类型数据,取值为 true 或 false,分别表示 WallOrRoad 对象是否为"路"、"墙"、"入口"或"出口"。
- point 是 MazePoint 类型对象,用来确定 WallOrRoad 对象在 Maze 所创建的迷宫容器中的位置。
- wallImage 和 roadImage 是 File 类型的对象,用来确定 WallOrRoad 对象上所绘制的图像文件。
- tool 是 Toolkit 对象,负责创建 Image 对象。

(2) 方法

- WallOrRoad()是构造方法,负责完成 WallOrRoad 对象的初始化。
- WallOrRoad 对象调用 setIsEnter(boolean)方法设置自己是否为迷宫的"入口"。

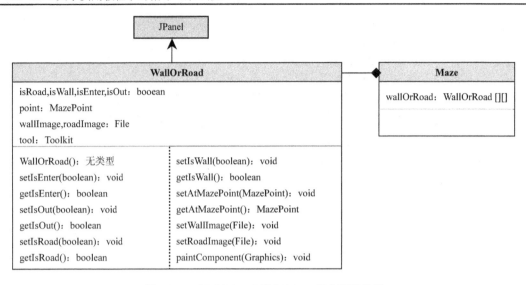

图 14-10　WallOrRoad 类与 Maze 类之间的关系

- WallOrRoad 对象调用 getIsEnter()方法判断自己是否是迷宫的"入口"，即返回 isEnter 属性的值。
- WallOrRoad 对象调用 setIsOut (boolean) 方法设置自己是否为迷宫"出口"。
- WallOrRoad 对象调用 getIsOut()方法判断自己是否迷为宫"出口"，即返回 isOut 属性的值。
- WallOrRoad 对象调用 setIsRoad(boolean)方法设置自己是否为迷宫中的"路"。
- WallOrRoad 对象调用 getIsRoad()方法判断自己是否为迷宫中的"路"，即返回 isRoad 属性的值。
- WallOrRoad 对象调用 setIsWall(boolean)方法设置自己是否为迷宫中的"墙"。
- WallOrRoad 对象调用 getIsWall()方法判断自己是否为迷宫中的"墙"，即返回 isWall 属性的值。
- WallOrRoad 对象调用 setAtMazePoint(MazePoint)方法设置自己所在的"点"，即设置 point 对象的引用。
- WallOrRoad 对象调用 getAtMazePoint()方法返回自己所在的"点"，即返回 point 对象的引用。
- WallOrRoad 对象调用 setWallImage(File)方法设置 wallImage 文件对象。
- WallOrRoad 对象调用 setRoadImage (File)方法可以设置 roadImage 文件对象。
- WallOrRoad 对象调用 paintComponent(Graphics)方法绘制图像，即绘制 tool 对象根据 roadImage 或 wallImage 文件所获得的 Image 对象。

3．代码（WallOrRoad.java）

```
import javax.swing.*;
import java.awt.*;
import javax.swing.border.*;
import java.io.*;
public class WallOrRoad extends JPanel{
    boolean isRoad,isWall,isEnter,isOut;
    MazePoint point;
    File wallImage,roadImage;
```

```java
    Toolkit tool;
    WallOrRoad(){
        tool=getToolkit();
    }
    public void setIsEnter(boolean boo){
        isEnter=boo;
        if(isEnter==true){
            add(new JLabel("入口"));
        }
    }
    public boolean getIsEnter(){
        return isEnter;
    }
    public void setIsOut(boolean boo){
        isOut=boo;
        if(isOut==true){
            add(new JLabel("出口"));
        }
    }
    public boolean getIsOut(){
        return isOut;
    }
    public void setIsRoad(boolean boo){
        isRoad=boo;
        if(isRoad==true){
            setBorder(null);
        }
    }
    public boolean getIsRoad(){
        return isRoad;
    }
    public void setIsWall(boolean boo){
        isWall=boo;
        if(isWall==true){
            setBorder(new SoftBevelBorder(BevelBorder.RAISED));
        }
    }
    public boolean getIsWall(){
        return isWall;
    }
    public void setAtMazePoint(MazePoint p){
        point=p;
    }
    public MazePoint getAtMazePoint(){
        return point;
    }
    public void setWallImage(File f){
```

```
            wallImage=f;
        }
        public void setRoadImage(File f){
            roadImage=f;
        }
        public void paintComponent(Graphics g){
            super.paintComponent(g);
            int w=getBounds().width;
            int h=getBounds().height;
            try{
                if(isRoad==true){
                    Image image=tool.getImage(roadImage.toURI().toURL());
                    g.drawImage(image,0,0,w,h,this);
                }
                else if(isWall==true){
                    Image image=tool.getImage(wallImage.toURI().toURL());
                    g.drawImage(image,0,0,w,h,this);
                }
            }
            catch(Exception exp){ }
        }
    }
```

14.3.5 MazePoint 类

1. 效果图

MazePoint 创建的对象负责确定 WallOrRoad 对象和 PersonInMaze 对象在 Maze 容器中的所在位置。MazePoint 创建的对象没有可显示的效果图。

2. UML 图

MazePoint 创建的对象含有两个重要 int 类型数据，分别表示 Maze 容器坐标系中的 x 坐标和 y 坐标值，坐标系的原点是 Maze 容器的左上角，向右是 X 轴的正向，向下是 Y 轴的正向。MazePoint 类创建的对象是二维数组 mazePoint 的单元中的对象。MazePoint 型二维数组 mazePoint 是 Maze 容器的重要成员之一，其单元中的 MazePoint 对象用来确定 WallOrRoad 对象和 PersonInMaze 对象在 Maze 容器中的位置。MazePoint 类的主要成员变量和方法及与 Maze 类之间的组合关系如图 14-11 所示。

（1）成员变量

- x、y 是 MazePoint 对象中的两个 int 类型数据，分别表示容器坐标系中的 x 坐标和 y 坐标的值。
- haveWallOrRoad 是 boolean 类型数据，如果有 WallOrRoad 对象在该 MazePoint 对象上时，haveWallOrRoad 的值是 true，否则为 false。
- wallOrRoad 是 WallOrRoad 类声明的对象，用来存放一个 WallOrRoad 对象的引用，表明该 WallOrRoad 对象在当前 MazePoint 对象上。

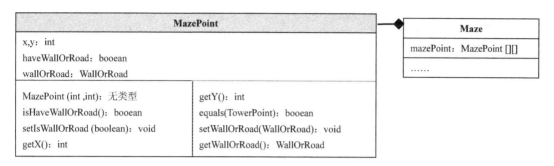

图 14-11　MazePoint 类的 UML 图

（2）方法
- MazePoint(int,int)是构造方法，用来创建 MazePoint 对象。
- MazePoint 对象调用 isHaveWallOrRoad()方法返回一个 boolean 类型数据，如果当前 MazePoint 对象上有 WallOrRoad 对象，isHaveWallOrRoad()方法返回 true，否则返回 false。
- MazePoint 对象调用 setIsWallOrRoad(boolean)方法可以根据参数的值设置当前 MazePoint 对象上是否有 WallOrRoad 对象。
- MazePoint 对象调用 getX()和 getY()方法，可以返回其中的 x 坐标和 y 坐标。
- MazePoint 对象调用 equals(MazePoint)方法，可以判断当前 MazePoint 对象是否与参数指定的 MazePoint 对象相同。
- Point 对象调用 setWallOrRoad(WallOrRoad)方法，将参数指定的 WallOrRoad 对象放置在当前 MazePoint 对象上。
- Point 对象调用 getWallOrRoad()方法，可以返回当前 MazePoint 对象上的 WallOrRoad 对象。

3. 代码（MazePoint.java）

```java
public class MazePoint{
    int x,y;
    boolean haveWallOrRoad;
    WallOrRoad wallOrRoad=null;
    public MazePoint(int x,int y){
        this.x=x;
        this.y=y;
    }
    public boolean isHaveWallOrRoad(){
        return haveWallOrRoad;
    }
    public void setIsWallOrRoad(boolean boo){
        haveWallOrRoad=boo;
    }
    public int getX(){
        return x;
    }
    public int getY(){
        return y;
    }
```

```
            public boolean equals(MazePoint p){
               if(p.getX()==this.getX()&&p.getY()==this.getY()){
                  return true;
               }
               else{
                  return false;
               }
            }
            public void setWallOrRoad(WallOrRoad obj){
               wallOrRoad=obj;
            }
            public WallOrRoad getWallOrRoad(){
               return wallOrRoad;
            }
         }
```

14.3.6　PersonInMaze 类

1．效果图

PersonInMaze 创建的对象效果如图 14-12 所示。

2．UML 图

PersonInMaze 类是 javax.swing 包中 JTextField 组件的一个子类，创建的对象 person 用来刻画迷宫中的"走迷宫者"，是 Maze 类的重要成员之一。PersonInMaze 类的主要成员变量和方法及与 Maze 类之间的关系如图 14-13 所示。

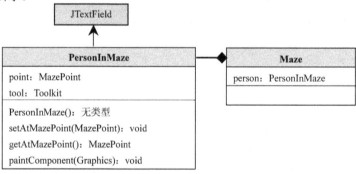

图14-12　PersonInMaze创建的对象　　　　图14-13　PersonInMaze类与Maze类之间的关系

（1）成员变量

- point 是 MazePoint 类型对象，用来确定 PersonInMaze 对象在 Maze 所创建的迷宫容器中的位置。
- tool 是 Toolkit 对象，用来获得 Image 对象。

（2）方法

- MazePoint()是构造方法，负责完成 MazePoint 对象的初始化。
- PersonInMaze 对象调用 setAtMazePoint(MazePoint)方法用来设置自己所在的"点"，即设置属性 point 的对象引用。

- PersonInMaze 对象调用 getAtMazePoint()方法返回自己所在的"点",即返回 point 对象的引用。
- PersonInMaze 对象调用 paintComponent(Graphics)方法绘制图像,即绘制 tool 对象所获得的 Image 对象。

3. 代码（PersonInMaze.java）

```java
import javax.swing.*;
import java.awt.*;
public class PersonInMaze extends JTextField{
    MazePoint point;
    Toolkit tool;
    PersonInMaze(){
        tool=getToolkit();
        setEditable(false);
        setBorder(null);
        setOpaque(false);
        setToolTipText("单击我,然后按键盘方向键");
    }
    public void setAtMazePoint(MazePoint p){
        point=p;
    }
    public MazePoint getAtMazePoint(){
        return point;
    }
    public void paintComponent(Graphics g){
        super.paintComponent(g);
        int w=getBounds().width;
        int h=getBounds().height;
        Image image=tool.getImage("person.gif");
        g.drawImage(image,0,0,w,h,this);
    }
}
```

14.3.7 HandleMove 类

1. 效果图

HandleMove 创建的对象效果如图 14-14 所示。

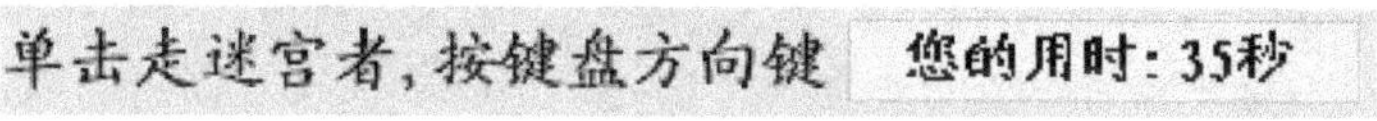

图 14-14　HandleMove 创建的对象

2. UML 图

HandleMove 类是 javax.swing 包中 JPanel 容器的一个子类,同时实现了 KeyListener 和 ActionListener 接口,创建的对象 handleMove 是 Maze 类的成员之一,负责监视 Maze 类中的 person 对象上的键盘事件。当用户用单击 person 对象,然后按键盘上的方向键时,handleMove 对象负责给出移动 person 对象的有关算法,并显示用户的用时。HandleMove 类的主要成员变量、方法及与 Maze 类之间的关系如图 14-15 所示。

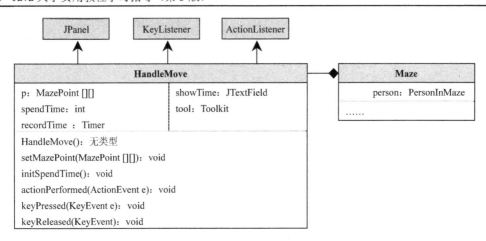

图 14-15　HandleMove 类与 Maze 类之间的关系

（1）成员变量
- p 是 MazePoint 类型的二维数组，用来存放 Maze 容器中 mazePoint 数组的引用。
- spendTime 用来记录用户移动"走迷宫者"，即移动 person 对象所用时间。
- recordTime 是计时器。
- showTime 是 JTextField 创建的文本框，负责显示用时。
- tool 是 Toolkit 对象，当移动 person 对象时，tool 对象负责调用 beep()方法发出"嘟嘟"声。

（2）方法
- HandleMove()是构造方法，负责创建 handleMove 对象。
- handleMove 对象调用 setMazePoint(MazePoint[][])方法，将 Maze 容器的 mazePoint 对象的引用传递给该对象中的 p，以便 handleMove 移动 Maze 容器中的 person 对象。
- handleMove 对象调用 initSpendTime()方法，将 spendTime 的值设置为 0。
- keyPressed(KeyEvent)方法是 HandleMove 类实现的 KeyListener 接口中的方法。如果单击 person 对象，然后按键盘上的方向键，将触发 KeyEvent 键盘事件，handleMove 对象将调用 keyPressed(KeyEvent)方法对事件做出处理，即根据规则移动 person 对象。
- keyReleased(KeyEvent)方法是 HandleMove 类实现的 KeyListener 接口中的方法。如果单击方向键，将触发 KeyEvent 键盘事件，handleMove 对象将调用 keyReleased(KeyEvent)方法对事件做出处理，主要判断用户是否已经成功地将 person 对象移动到迷宫的出口处。
- actionPerformed(ActionEvent)是 HandleMove 类实现的 ActionListener 接口中的方法。我们使用 Timer 类的构造方法 Timer(int a, Object b)创建了计时器 recordTime，其中的参数 a 的单位是毫秒，确定计时器每隔 a 毫秒"震铃"一次，参数 b 是计时器的监视器。这里取 a=1000，监视器 b 是当前 handleMove 对象。计时器发生的震铃事件是 ActinEvent 事件，震铃事件发生时，监视器会监视到这个事件，就调用 actionPerformed ActionEvent)方法。当震铃每隔 1000 毫秒发生一次时，方法 actionPerformed(ActionEvent)就被执行一次，所执行的操作就是显示用户当前的用时。

3. 代码（HandleMove.java）

```
import java.awt.event.*;
import java.awt.*;
```

```java
import javax.swing.*;
public class HandleMove extends JPanel implements KeyListener,ActionListener{
    MazePoint [][] p;
    int spendTime=0;
    javax.swing.Timer recordTime;
    JTextField showTime;
    Toolkit tool;
    HandleMove(){
        recordTime=new javax.swing.Timer(1000,this);
        showTime=new JTextField(16);
        tool=getToolkit();
        showTime.setEditable(false);
        showTime.setHorizontalAlignment(JTextField.CENTER);
        showTime.setFont(new Font("楷体_GB2312",Font.BOLD,16));
        JLabel hitMess=new JLabel("单击走迷宫者,按键盘方向键",JLabel.CENTER);
        hitMess.setFont(new Font("楷体_GB2312",Font.BOLD,18));
        add(hitMess);
        add(showTime);
        setBackground(Color.cyan);
    }
    public void setMazePoint(MazePoint [][] point){
        p=point;
    }
    public void initSpendTime(){
        recordTime.stop();
        spendTime=0;
        showTime.setText(null);
    }
    public void keyPressed(KeyEvent e){
        recordTime.start();
        PersonInMaze person=null;
        person=(PersonInMaze)e.getSource();
        int m=-1,n=-1;
        MazePoint startPoint=person.getAtMazePoint();
        for(int i=0;i<p.length;i++){
            for(int j=0;j<p[i].length;j++){
                if(startPoint.equals(p[i][j])){
                    m=i;
                    n=j;
                    break;
                }
            }
        }
        if(e.getKeyCode()==KeyEvent.VK_UP){
            int k=Math.max(m-1,0);
            if(p[k][n].getWallOrRoad().getIsRoad()){
                tool.beep();                                //发出嘟的一声
```

```java
                    person.setAtMazePoint(p[k][n]);
                    person.setLocation(p[k][n].getX(),p[k][n].getY());
                }
            }
            else if(e.getKeyCode()==KeyEvent.VK_DOWN){
                int k=Math.min(m+1,p.length-1);
                if(p[k][n].getWallOrRoad().getIsRoad()) {
                    tool.beep();
                    person.setAtMazePoint(p[k][n]);
                    person.setLocation(p[k][n].getX(),p[k][n].getY());
                }
            }
            else if(e.getKeyCode()==KeyEvent.VK_LEFT){
                int k=Math.max(n-1,0);
                if(p[m][k].getWallOrRoad().getIsRoad()){
                    tool.beep();
                    person.setAtMazePoint(p[m][k]);
                    person.setLocation(p[m][k].getX(),p[m][k].getY());
                }
            }
            else if(e.getKeyCode()==KeyEvent.VK_RIGHT){
                int k=Math.min(n+1,p[0].length-1);
                if(p[m][k].getWallOrRoad().getIsRoad()){
                    tool.beep();
                    person.setAtMazePoint(p[m][k]);
                    person.setLocation(p[m][k].getX(),p[m][k].getY());
                }
            }
        }
        public void actionPerformed(ActionEvent e){
            spendTime++;
            showTime.setText("您的用时:"+ spendTime+ "秒");
        }
        public void keyReleased(KeyEvent e){
            PersonInMaze person=(PersonInMaze)e.getSource();
            int m=-1,n=-1;
            MazePoint endPoint=person.getAtMazePoint();
            if(endPoint.getWallOrRoad().getIsOut()){
                recordTime.stop();
                JOptionPane.showMessageDialog(this,"您成功了!","消息框",
                                      JOptionPane.INFORMATION_MESSAGE);
            }
        }
        public void keyTyped(KeyEvent e){ }
}
```

14.3.8 所需图像

准备名为"road.jpg"、"wall.jpg"和"person.jpg"的图像。其中,"road.jpg"和"wall.jpg"是迷宫中"路"和"墙"的默认图像,"person.jpg"是"走迷宫者"的默认图像。

14.4 代码调试

将 14.3.1 节中的迷宫文件、14.3.2 节~14.3.7 节中的 6 个 Java 源文件、14.3.8 节中的 3 幅图像保存到同一目录中,如 D:\ch14 中。分别编译这 6 个 Java 源文件,或用 "javac *.java" 命令编译全部的源文件,然后运行主类,即运行 MazeWindow 类。

14.5 软件发布

可以使用 jar.exe 命令制作 JAR 文件来发布软件。

首先用文本编辑器,如 Windows 下的记事本,编写一个清单文件。

mymoon.mf

```
Manifest-Version: 1.0
Main-Class: MazeWindow
Created-By: 1.2(Sun Microsystems Inc.)
```

将 mymoon.mf 保存到 D:\ch14 中,即与应用程序所用的字节码文件保存在相同的目录中。

注意:清单文件中的 Manifest-Version 与 1.0 之间、Main-Class 与主类 MazeWindow 之间、Created-By 与 1.2 之间必须有且只有一个空格。

然后生成 JAR 文件:

```
D:\ch14\jar cfm Maze.jar mymoon.mf *.class
```

其中,参数 c 表示要生成一个新的 JAR 文件,f 表示要生成的 JAR 文件的名字,m 表示清单文件的名字。

现在就可以将 Maze.jar、road.jpg、wall.jpg 和 person.jpg 复制到任何一个安装了 Java 运行环境(版本号需高于 1.2 的计算机上)的计算机上,双击该文件的图标就可以运行该软件。

14.6 实验后的练习

改进本章的"走迷宫"程序,具体要求如下:

(1) 对相应的迷宫增加"英雄榜"功能。当用户成功将该迷宫中的走迷宫者从入口移动到出口后,如果成绩能排进前三名,就弹出一个对话框,将用户的成绩保存到"英雄榜"。

(2) 增加查看"英雄榜"的功能。

本程序使用 Toolkit 对象调用 beep()方法使得走迷宫者在移动过程中发出声音,但该声音单调无味。请改进程序,增加更加丰富的音乐效果,在 HandleMove 中增加播放音乐的功能模块,用户成功移动走迷宫者后,程序播放简短的一声音乐。用 Java 可以编写播放 AU、AIFF、WAV、MID、RFM 格式的音频。假设音频文件 hello.au 位于应用程序当前目录中,有关播放音乐的知识如下:

① 创建 File 对象（File 类属于 java.io 包）
    File musicFile=new File("hello.au");
② 获取 URI 对象（URI 类属于 java.net 包）
    URI uri=musicFile.toURI();
③ 获取 URL 对象（URL 类属于 java.net 包）
    URI url=uri.toURI();
④ 创建音频对象(AudioClip 和 Applet 类属于 java.applet 包)
    AudioClip clip=Applet.newAudioClip(url);
⑤ 播放、循环、停止
    clip.play()　　　　　　　　　　//开始播放
    clip.loop()　　　　　　　　　　//循环播放
    clip.stop()　　　　　　　　　　//停止播放

（4）改进程序，允许用户更改或者自定义迷宫中"墙"和"路"上的图像。

（5）准备一幅图像，名为"pig.jpg"。规定一个时间上限，如 3 分钟。如果用时超过所规定的上限，"走迷宫者"上的图像变为 pig.jpg。